INTEGRATED PROJECT DELIVERY
An Action Guide for Leaders

INTRODUCTION

This guide is the result of a collaborative writing process between five IPD subject-matter experts that made up the guide's core team. Much of the content found in this guide derived from the conversations that took place at the IPD Advisory Council workshop on May 2, 2017, at the Center for Education and Research in Construction of the Department of Construction Management at the University of Washington. At this workshop, the IPD Advisory Council—a group of twenty IPD-experienced industry professionals, representing public and private owners, contractors, architects, and trades from across the US and Canada—gathered with our subject-matter experts to discuss choosing IPD and establishing goals; actions for when things go wrong; project processes and tools; and legal, commercial, management, and team-culture strategies and best practices. These workshop conversations were recorded, transcribed, and then expanded on by the core team. A completed draft of the guide was then peer reviewed by a group of eighteen industry professionals and academic experts. Following the peer review, the core team integrated the reviewers' comments into the guide, creating the guide you see here. Understanding the dynamic nature of IPD and the building industry, this guide represents IPD as it is currently practiced.

SPONSORS

Charles Pankow Foundation
Center for Innovation in the Design and Construction Industry (CIDCI)
Integrated Project Delivery Alliance (IPDA)

ADDITIONAL FUNDING PROVIDED BY...

Array Architects
Boldt
Cammisa + Wipf
CH2M (now Jacobs)
Chandos Construction
Charles Pankow Builders
Clark Construction
DLR Group
DPR Construction
Ferguson Corporation
Gilbane
Gillam Group
Group 2 Architecture Interior Design
Procter & Gamble
Robins & Morton
Rosendin Electric
Southland Industries
Whiting-Turner

(see also p.152 / inside back cover)

CORE GROUP

Markku Allison, Chandos Construction
Howard Ashcraft, Hanson Bridgett LLP
Renée Cheng, University of Minnesota
Sue Klawans
James Pease, Lean IPD

EDITORS

Renée Cheng
Laura Osburn
Linda Lee (copy editor)

PROJECT MANAGER

Laura Osburn, University of Washington

GRAPHIC DESIGN AND INFOGRAPHICS

MGMT. design

IPD ADVISORY COUNCIL

Jack Avery, Sellen Construction, *United States*
James Bedrick, AEC Process Engineering, *United States*
Dan Borton, Amgen, *United States*
Carl Davis, Array Architects , *United States*
Stuart Eckblad, University of California,
  San Francisco Medical Center, *United States*
Dominic Esparza, Southland Industries, *United States*
Clay Goser, University of Chicago, *United States*
Michael Guglielmo, Brown University, *United States*
Jen Hancock, Chandos Construction, *Canada*
Ken Jaeger, Red Deer Catholic Regional Schools, *Canada*
Lydia Knowles, Haworth, *United States*
Jason Martin, Boldt, *United States*
Michael McCormick, University of Washington,
  United States
Ron Migliori, Buehler & Buehler Structural Engineers,
  United States
Richard Neal, Ferguson Corporation, *Canada*
Rory Picklyk, Stantec, *Canada*
Christian Pikel, Universal Health Services, *United States*
Dean Reed, DPR Construction, *United States*
Pamela Touschner, DLR Group, *United States*
Craig Webber, Group 2 Architects, *Canada*

PEER REVIEWERS

James Bedrick, AEC Process Engineering
Phillip Bernstein, Yale University
Dan Borton, Amgen
Carrie Sturts Dossick, University of Washington
Kelly Fawcett, Canadian Nuclear Laboratories
Jen Hancock, Chandos Construction
Vicky Hooper, HGA Architects
Steven Innes, Canadian Nuclear Laboratories
Jason Martin, Boldt
Ron Migliori, Buehler & Buehler Structural Engineers
Bob Minutoli, Whiting-Turner
Christian Pikel, ReAlignment Group (formerly with
  Universal Health Services)
Dean Reed, DPR Construction
Craig Russell, Walt Disney Imagineering
Sheryl Staub-French, University of British Columbia
John Strickland, CH2M (now Jacobs)
Pamela Touschner, DLR Group
Craig Webber, Group 2 Architecture Interior Design

WE WOULD ALSO LIKE TO THANK...

Paulo Napolitano and Herrero Builders
Patrick Lencioni and The Table Group
Cara Carmichael, Rocky Mountain Institute
Kent Hedges, Universal Health Systems

CONTENTS

FROM BEGINNING TO END: ONGOING CONSIDERATIONS 39

EARLY WORK: PROCESSES AND TOOLS 61

LATER WORK: MAINTAINING MOMENTUM 91

WHAT GOES WRONG—AND WHAT CAN WE DO ABOUT IT? 105

MORE RESOURCES 111

GLOSSARY 113

APPENDICES 119

Columns (phases): PRE-VALIDATION · VALIDATION · DETAILED DESIGN AND IMPLEMENTATION DOCUMENTS

PATH TO CONTRACT

Owner Alignment

Team Selection

The Contract Workshop/Team Alignment

ONGOING CONSIDERATIONS

Team Management Building and Managing a Successful IPD Team

Financial Organization and Financial Monitoring

Lean Thinking

EARLY WORK

Validation: Go/No Go

Target Value Design

Co-location in a Big Room

Design Management

Prefabrication

Integrating Project Information Using Building Information Modeling

Risk Management

Project Dashboards

LATER WORK

Team Maintenance

WHAT GOES WRONG

WHEN TO FOCUS?

During the chronology of a typical IPD project, there are times when the team needs to invest focused effort. Project phases are indicated across the top of the page; horizontal bars correspond to topics or areas of effort; while most of these topics warrant consistent reflection across the project, denser/thicker bars indicate times at which special attention might be paid. Often, more intense times are the start or end of particular tasks or phases.

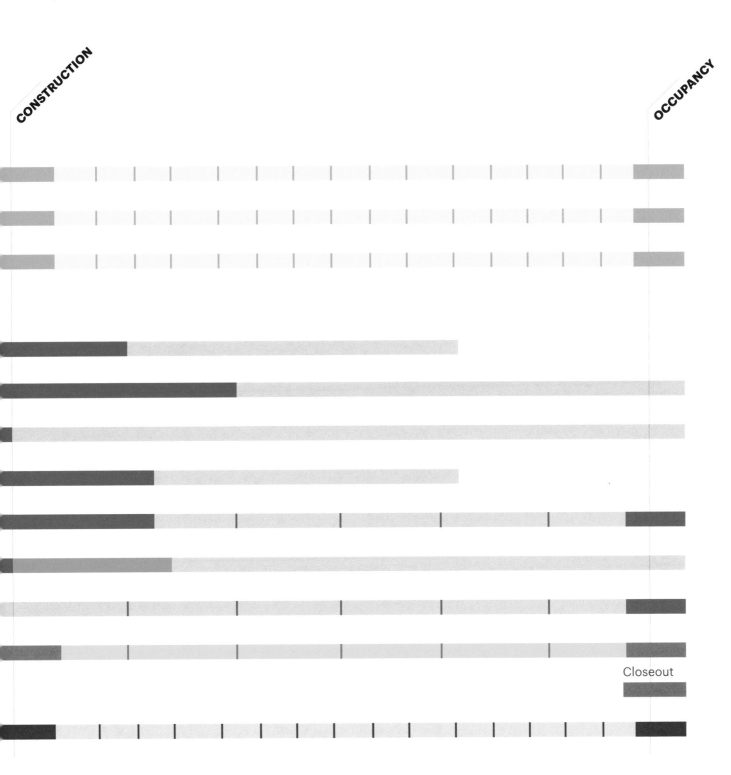

PREFACE
How to Use This Guide

___

What is the Purpose of This Guide?

Over the past decade, Integrated Project Delivery (IPD) has emerged as an important alternative to traditional forms of project delivery. Its adherents have reported improvements in cost, schedule, and quality, achieved in an atmosphere that is less adversarial and more collaborative. Moreover, research on IPD, as well as detailed case studies, have documented this project delivery system's value proposition.

If you are unsure of the value proposition, we suggest you reach out to owners or other participants who have used IPD and ask them about their experiences. In addition, you should review the materials referenced in the More Resources section of the guide to understand the essential principles and requirements of IPD. However, if you are reading this guide, we assume that you have decided to give IPD a try and are now wondering, what do I do next? In the following pages, we have attempted to answer that and related questions concerning the commencement and execution of IPD projects. This is a practical, not a theoretical guide. It is based on the experience of the core group, the advisory council, and the peer reviewers and is intended to help you understand what you will need to do to have a successful IPD project.

The guide is not a replacement for consulting with IPD advisors and coaches. Almost all parties and teams need assistance, at least with their first project. The guide will, however, provide you with a good understanding of whether IPD fits your project and organization, what needs to be done, your role in the process, what problems can occur, and counter-measures

you can employ. It is also not a manual on lean principles and processes, target value design (TVD), or virtual design and construction, although these will be touched on in the manual. Detailed information regarding these topics is well presented in the reference materials and should be adapted to the specific project.

How is This Guide Organized?

The guide is organized in a roughly chronological order and divided into specific blocks to allow the reader to immediately access information relevant to their project concerns. Most readers will benefit from reading from the beginning through the end. Others may need specific information or will want to review specific sections during project execution. The guide can be used in any of these ways.

Care was taken in selecting the terminology of each of the guide's sections. In particular, our guide's core group adopted the terms "early work" and "later work" to indicate specific periods in a project timeline while avoiding terminology (such as pre-design or pre-construction) that may have different meanings for different stakeholder types. These terms also avoid the usual conception of projects having distinct design and construction phases. In IPD, owners, designers, and contractors come together early and often in the project so distinguishing between phases is not meaningful.

The guide is divided into five sections, each of which answers fundamental questions that may arise during specific periods in a project's life cycle. Some concepts are applicable throughout the project and have been grouped together or are discussed in several sections in ways appropriate to that stage of the project.

> **Path to Contract** provides a road map for creating internal alignment within your organization, assembling and aligning the project team, and creating an appropriate agreement that binds them together. We have found that a robust process leads to better alignment, fewer problems, and a higher probability of success.

> **From Beginning to End** describes management, financial, and lean considerations that are important throughout the process. You should address these issues at commencement and continuously throughout the project.

> **Early Work** focuses on the key tasks in organizing and commencing a project, including validation, Target Value Design, co-location, design management, prefabrication, Building Information Modeling (BIM), managing risk, and managing metrics. In IPD, the amount of early work is significantly increased because addressing these issues and creating effective processes leads to the greatest gains.

> **Later Work** continues the concepts developed in Early Work and discusses maintaining momentum on the project. In our experience, if the Early Work is done well, the Later Work will naturally follow. For this reason, we have focused more guide material on Early Work than Later Work.

> **What Goes Wrong—and What Can We Do About It** deals with the unfortunate reality that not everything works perfectly every time. On any project, things will go wrong. But IPD provides you with an engaged and motivated team to solve problems. It helps, however, to be able to recognize potential problems before they occur. In this section, we have listed the most common failings on IPD projects and have recommended countermeasures to regain desired outcomes.

At the end of these sections are additional materials that supplement the information in the guide.

> **More Resources** and **Glossary** contain materials that provide additional detail and support for concepts and terms in the guide.

> The **Appendices** of templates and samples provide IPD tools and artifacts that can help you plan and execute your own project.

The guide is more deeply focused on initial project organization than on execution during the construction phase. This is not because the construction phase is less important. Rather, the experience of IPD teams has been that a project that is properly aligned, validated, and managed from inception is likely to be well managed during construction. Thus, our emphasis is on creating the right collaborative environment and engaging in thorough planning of processes as well as the project. If these are done well, we are confident the IPD team will build on this framework and execute construction well.

IPD can provide superior outcomes over a wide range of project types, but it is not appropriate for every project or owner.

IS IPD RIGHT FOR YOU AND YOUR PROJECT?

IPD can provide superior outcomes over a wide range of project types, but it is not appropriate for every project or owner. Because IPD is cost-based and team managed, it inherently requires greater leadership and administrative effort than a lump sum, hierarchical project. This effort is amply repaid on most projects—but not all. Before commencing on your first IPD project, evaluate the project and assess your organization's culture, capabilities, and resolve. Moreover, a good IPD owner does more than just participate: he or she is a facilitative leader that models collaborative behavior. Therefore, honestly evaluate your willingness (and ability) to embrace the changes in process and procedure that drive IPD's benefits.

Is Your Project Right for IPD?

Table 1 lists project characteristics relevant to matching IPD to a specific project. There is no fixed formula for knowing if IPD is right, but if an honest evaluation of the factors for Ambition, Stressors, Level of Clarity, Probability of Change, and Complexity are predominantly "Low," then another project delivery approach may be more appropriate.

Other factors to consider:

Context. The characteristics of Project Size and Project Status may also bear on the decision to use IPD, but these must be considered in context. If a team has never undertaken an IPD project, it must invest time in contract creation, organization, and training. Smaller projects may not be able to absorb these additional costs and additional level of effort. However, a team with extensive IPD experience can efficiently execute smaller projects. Although it is best to start IPD when a project commences, and certainly before design is complete, in some projects there may still be adequate benefits of converting a project to IPD later in the schedule. Another factor to consider is the ability to use IPD on a portfolio of projects. This creates an opportunity to transfer learnings from one project to subsequent projects, developing greater skill and success in project execution. Finally, the sophistication of trade contractors and consultants in the local market may affect a decision to use IPD. Although many IPD projects have been successfully performed with participants who have never done IPD projects, prior experience is beneficial.

Ratio of Project Size to Team IPD Experience. Because IPD is a cost-based system, it requires more accounting effort than a lump sum contract, and this effort may not be justified on a smaller project. Experienced IPD teams have successfully executed projects as small as $1.5M, and some owners use IPD on their projects starting at $5M. However, if the team is not experienced with IPD, a larger project is necessary to absorb the additional training and organizational effort required. If the majority of the team is inexperienced, project minimums are often closer to $15M in order to absorb first time coaching and training expenditures. Project minimums may be even higher, if none of the parties, including the owner, have IPD experience.

Phase of Design and Opportunity for Innovation. The longer a project has been in design, the less opportunity for innovation and target value design (TVD). Moreover,

TABLE 1: PROJECT CHARACTERISTICS

PROJECT CHARACTERISTIC		HIGH	LOW
Level of Ambition	Technical Innovation	○	○
	Creative Innovation	○	○
	Other Areas of Innovation	○	○
	High Sustainability Goals	○	○
Stressors	High Value to Budget	○	○
	Challenging Schedule	○	○
Level of Clarity[1]	Current Scope Development	○	○
	Expected Time for Future Scope Development	○	○
Probability of Change	Expected Change in Building Technology	○	○
	Expected Change in Business Case	○	○
	Expected Stakeholder / Public Driven Change	○	○
Complexity of Interaction	Level of Interdependency of Systems	○	○
	Level of Interdependency of Participants	○	○

the project participants may have "cemented" their working relationships. If IPD begins well after conceptual design, the owner should expect less to be gained, although improvement is still possible.

Is Your Organization Right for IPD?

Just as every project is not right for IPD, IPD is not right for every person or organization. However, unlike projects with relatively fixed characteristics, individuals and organizations can change. And the actions of individuals over time, change organizations. The key determinant is not whether your organization is a perfect fit for IPD but whether your organization is able and committed to make the changes necessary for successful IPD implementation. If the answer is "not quite yet," your organization may still be a candidate because IPD is a journey and the "fit" between IPD and your organization will evolve over time. This evolution is particularly important for owners that repeatedly build. They should focus on creating a continuously improving and reliable system of project management

rather than immediate project-specific benefits. If your organization is not ready for a complete transformation, it can consider approaching IPD on a pilot project, using a select team that can demonstrate the effectiveness of IPD for your organization. As one experienced owner commented, "Sometimes, the best way to begin an IPD transition is just start."

The Owner's Organization

The owner's role is very significant in an IPD project. The owner defines its goals, provides direction, and models transparency and collaboration to the entire project team. Moreover, you cannot remove the owner's organization from the project as you can with other participants. Although the attributes discussed below apply to all IPD participants, they are especially important for the owner.

---

1 IPD is a good choice when managing projects with scopes that are not initially clear, but it will require a more extended validation period before setting targets

Many designers who have worked in an IPD environment report that it is richer and more invigorating than traditional practice and that the relationship and trust built between designers and contractors allows the designers to focus on the creative aspects of their practice.

The perfect owner (we know that this is an abstraction) has the following characteristics:

> **Engaged**

IPD is not a spectator sport. The owner's personnel must engage in the project management team, senior management team, and must participate in project implementation teams that perform the actual work. If you are an owner, evaluate your organization: do you have the capacity to engage deeply in the project? Engagement requires effort[2], but it is also an opportunity because it allows the owner to directly influence the project's outcome. Moreover, by being directly engaged, the owner eliminates the propose/review/approve cycles that can result in delay and rework. The engaged owner is always aware of project status and progress and knows, long before project completion, exactly what they will be receiving. While the owner can supplement their team with consultants to handle various administrative tasks, the owner cannot delegate its decision-making responsibility.

> **Collaborative**

If an owner wants to reap the rewards of collaboration, the owner must also be collaborative. If the project team senses that the owner is authoritarian and only concerned with their own issues, the team members will feel free to place their own company's needs above the project's needs. This is incompatible with the IPD "project first" sensibility. Honesty and transparency by the owner is also paramount. Many owners believe that they should withhold key information—for example, budgets and contingency amounts—to get the most aggressive pricing from the project team. In IPD interaction, a team is most innovative when they fully understand the owner's business case and resources. Finally, the owner also needs to be reliable, meeting obligations, keeping promises, and making decisions based on the timing needs of the project team.

---

2 Opinions differ whether IPD requires more owner effort than other project delivery types. Some owners believe it is less, some more, and some the same. However, there is general agreement that, with IPD, more work occurs during planning and that owners are less focused on addressing problems and claims and more focused on defining what they want to achieve and on working with the team to accomplish the goals.

The owner's project manager should be chosen with care. She or he should facilitate team performance and be mindful of team dynamics and the interests of others. Unfortunately, many experienced project managers have been scarred by prior traditional-project experiences and hesitant to engage in trust-based relationships required with designers and builders in IPD. Traditionally, experienced managers were trained to use command-and-control approaches to management: issuing orders, directing tasks, criticizing performance, and enforcing penalties for missed deadlines. Unless they learn to manage in a collaborative enterprise, traditional project managers may cripple the IPD project. Seriously consider whether you and your organization have the willingness and skills to participate in, and support, a high-performance team. Skills can be augmented by engaging coaches or consultants, but willingness must exist within the organization.

> **Competent**

It is difficult for an owner to engage with the project team if owner requests are not well-informed and realistic. The owner's project manager needs to have at least a general understanding of the design and construction process to participate in project decisions. To some degree, a less experienced owner can compensate by adding consultants as trusted advisors, provided they have the right disposition and experience for the task. Many traditional program managers and independent consultants view their role as "protecting" the owner from "deceitful" contractors and designers and may have a belligerent style. Choose consultants who believe their best role is to enable team performance and monitor their performance to see that they are not reverting to a reward-and-punishment style of management.

> **Committed**

Because IPD projects often challenge existing practices and engage in new ways of determining value and innovation, project teams work through many problems. Inevitably, they will make mistakes. The first time this happens, the owner may be tempted to abandon the IPD process. This is a critical moment. The owner can vacillate (which will deflate the project team) or demonstrate commitment to the process. Displaying commitment will reap rewards because it shows the project team that the owner will support them when they explore options and seek innovative solutions. One experienced project manager commented that it was very helpful to know that management was solidly behind IPD. Others noted that having a highly placed IPD champion allows everyone to concentrate on delivering the project instead of defending their careers.

We know that very few organizations will perfectly match the profile described above—if they did, they would probably be using IPD already. However, it is important to understand that change is difficult and if an organization

WHAT ARE THE ADVANTAGES OF DIFFERENT DELIVERY METHODS ON THIS PROJECT?

To make sure IPD is right for this project, you can build a Choosing by Advantages (CBA) matrix to analyze the advantages of different delivery systems. After establishing your goals, build a CBA matrix that includes input from users, builders, and designers. Then vet the process to determine which delivery system is best for your project.

is inherently mistrustful, hierarchical, and accusatory, it will be hard for it to participate in a high-performance collaborative environment. Moreover, in early stages of an IPD transition, it may be necessary to protect the project team from internal forces until the organizational culture has adapted to IPD. Some owners have described an "IPD bubble" for pilot projects, creating a zone of experimentation while overall organizational culture shifts to accept IPD as a long-term strategy. The owner has a role in developing IPD managers and leaders and protecting the project team during this organizational transformation.

We have primarily focused on the owner's organization as it is at the center of the IPD project and will have the greatest effect on project culture. While most of the points regarding the owner's organization are valid for other participants as well, there are also a few considerations particular to specific disciplines.

Designers

Designers, more than other IPD participants, have traditionally worked from their offices, collaborating mostly with their own colleagues. Customarily, they gather information from the client, generate concepts, and when they are adequately developed, present them for review and critique. This process iterates until the client is satisfied. In contrast, IPD projects have a continuous design flow with all participants able to view the design as it is developing and provide frequent commentary and suggestions. For example, designers need to accept cost feedback based on a preliminary design that may come without a fully detailed estimate. Moreover, the entire process may take place in a co-located environment with continuous interaction among the entire project team. Many designers who have worked in an IPD environment report that it is richer and more invigorating than traditional practice and that the relationship and trust built between designers and contractors allows the designers to focus on the creative aspects of their practice. However, others—particularly those with little IPD experience—are disquieted by the open and fluid nature of IPD design and are worried that creativity will be compromised.

Because designers are also involved in IPD project management, they must understand the project from a broader perspective than design alone. This implies a higher level of understanding of project management and leadership, cost and schedule controls, and similar issues. The IPD designer must also be able to predict and manage their own budgets and deliverables in step with the work of the other project team members. Although this requires additional effort, designers who engage at the project level will gain a far deeper understanding of the project than those who remain isolated in their offices.

Designers, and particularly the consultants to the lead designers, must understand that co-location is a key IPD methodology and must be able to commit the time to work collaboratively with others, often at some distance from their offices. Good IPD teams recognize that this can be difficult for some participants and implement processes to allow for virtual connection.

GETTING THE MOST VALUE ON COMPLEX PROJECTS

"For the owner, before you make the decision to use IPD, you have to ask, 'what is the type and level of complexity of the project?' When we did a pilot project on a very complex project, but with a small dollar value, only $12 million, we asked how to get the most value out of that $12 million. If we had completed the project using conventional methods, we probably would have had only 75% of the scope of work completed because we would have drawn it the wrong way. Once we had the trade partners around the table, they were actually able to solve the challenge of working with a limited budget with the designers. Truly breaking down the silos."   —Owner

However, for at least part of the project, some physical co-location will be required. Designers need to embrace the opportunity, not fight it.

If you are a designer, ask yourself whether you willing to adopt new, more collaborative workflows, be physically co-located, and step up to a larger and more influential project role.

If you are a CM/GC, ask yourself whether your project leaders are capable of coaching, guiding, and facilitating, in addition to managing cost and schedule.

Construction Manager/ General Contractor

The IPD process requires project leadership and careful management of value. The construction manager/ general contractor (CM/GC) is central to these tasks and has a special role in IPD. The CM/GC must be skilled in conceptual cost development (based on design information not yet fully detailed) to provide feedback for TVD. Along with other IPD members, the CM/GC must be a facilitative leader focused on achieving the overarching project goals, not just focused solely on cost and schedule. To do this, the CM/GC must understand what the owner is trying to accomplish and the skills needed to energize the entire project team's talents to achieve these ends. This is a very different role from traditional project delivery: authoritarian command and control methodologies do not work in this new context. Additionally, the CM/GC must be just as transparent as all other parties and must view designers and trade contractors as equal participants in the process.

If you are a CM/GC, ask yourself whether your project leaders are capable of coaching, guiding, and facilitating, in addition to managing cost and schedule.

Trade Contractors

Trade contractors, some of whom might be trade partners, are expected to be directly involved in design development, rapid cost analysis, and constructability, as well as traditional tasks of procurement and construction. They will jointly be responsible for developing the schedule and determining construction sequences and methods. Although prior experience with design assist or design/build can be beneficial, most trade contractors report that IPD requires more services from them and has greater accounting requirements because of the increased detail in cost forecasting and reporting. They need to understand that they have a responsibility for the whole (at least with regard to their profit) and cannot just comply with the plans and specifications given to them. Moreover, they need to support the development of conceptual estimates, rather than estimating from finished designs, to enable TVD.

If you are a trade contractor, enjoy your enhanced project role but also assess whether you have the capabilities and staff required for IPD.

WHAT IS A PROJECT?

In a project with a high-level understanding, there is often an intersection of three spheres: financial goals, owner business objectives, and team culture.

Team culture includes the vast range of so-called soft elements, including shared values, alignment of goals, creation of psychological safety, mutual trust, and respect as well as lean thinking and a "project first" attitude. An IPD team invests a substantial amount of time and energy in establishing, supporting, and maintaining a strong team culture.

Owner business objectives are critically important for the IPD team, since a building project is the solution to the owner's needs, not an end in itself. Business objectives typically include meeting budget and schedule goals but go well beyond into areas of performance, brand identity, employee wellness, and societal impact.

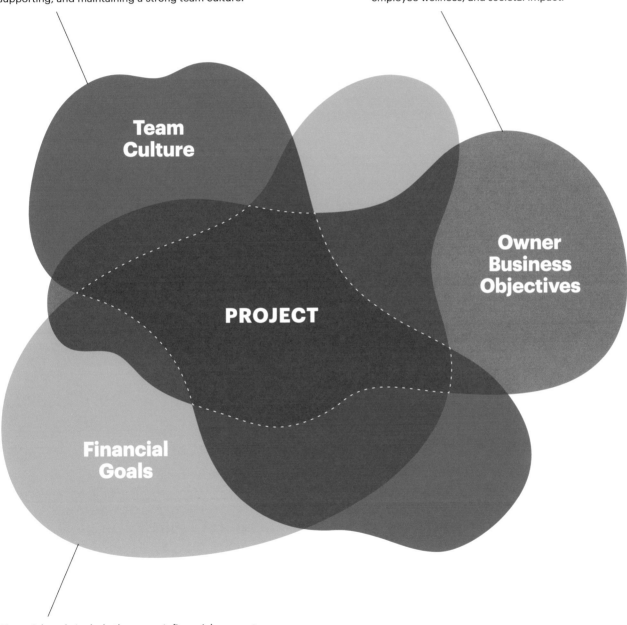

Financial goals include the owner's financial parameters and funding criteria. It also includes the financial incentives for the IPD team—taking into account how each member of the team structures profit, overhead, and labor. In many cases, the financial goals include nonmonetary or indirect value for the owner and team.

PRE-VALIDATION

VALIDATION

DETAILED DESIGN AND IMPLEMENTATION DOCUMENTS

PATH TO CONTRACT

Owner Alignment

Team Selection

The Contract Workshop/Team Alignment

ONGOING CONSIDERATIONS

Team Management Building and Managing a Successful IPD Team

Financial Organization and Financial Monitoring

Lean Thinking

EARLY WORK

Validation: Go/No Go

Target Value Design

Co-location in a Big Room

Design Management

Prefabrication

Integrating Project Information Using Building Information Modeling

Risk Management

Project Dashboards

LATER WORK

Team Maintenance

WHAT GOES WRONG

Closeout

PATH TO CONTRACT
Establishing Common Purpose

―――――

IPD is an intentional process driven by continuous improvement. From the very beginning, the IPD team needs to plan how to execute the project, put the plan in motion, evaluate the plan's effectiveness, and revise the plan for greater success. This process shifts effort to earlier stages of the project. The early, relatively high burn rate can create anxiety for team members new to IPD. While the early stages must be managed efficiently, the early burn rate is a necessary consequence of increased early planning, which will pay dividends as the project unfolds.

The process of negotiating and executing an IPD agreement should be a positive step that aligns the entire IPD team to the project goals and creates a commitment to execute the project collaboratively. It will require effort—and likely assistance—to manage properly, but it will provide benefits that significantly outweigh the cost and effort.

PATH TO CONTRACT

Each owner and each project will have its own context and set of challenges. Therefore, the process should be adjusted to meet your specific requirements. The general framework shown below is appropriate for many projects and can serve as a starting point toward a more tailored process. In general, the process begins with owner alignment and then proceeds to team selection, team alignment/contract workshop, and then post-workshop tasks. In some cases, the steps in the process may be overlapped or run in parallel with work done for validation.

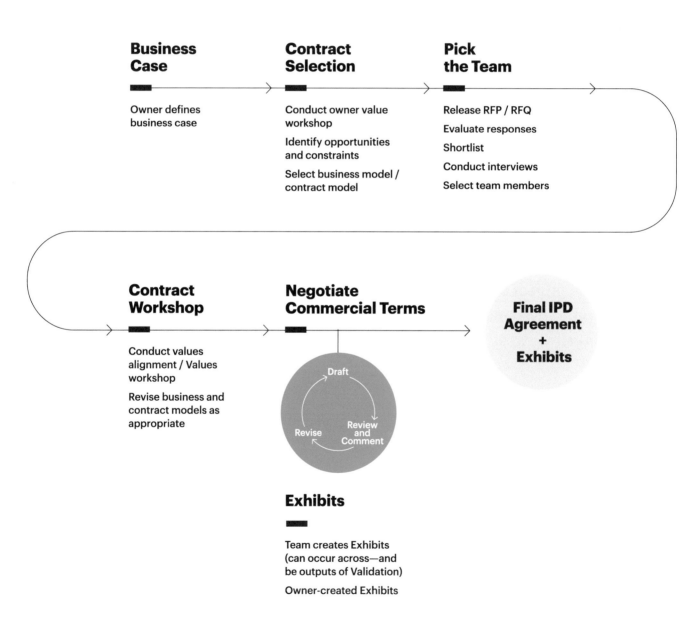

Business Case

Owner defines business case

Contract Selection

Conduct owner value workshop

Identify opportunities and constraints

Select business model / contract model

Pick the Team

Release RFP / RFQ

Evaluate responses

Shortlist

Conduct interviews

Select team members

Contract Workshop

Conduct values alignment / Values workshop

Revise business and contract models as appropriate

Negotiate Commercial Terms

Draft

Review and Comment

Revise

Final IPD Agreement + Exhibits

Exhibits

Team creates Exhibits (can occur across—and be outputs of Validation)

Owner-created Exhibits

OWNER ALIGNMENT
Is Everyone on the Same Page?

What is owner alignment?

Although IPD is a collaborative project delivery method, the owner has a special responsibility to set and align the expectations of all owner stakeholders. In parallel, owner alignment also includes ensuring that its management and stakeholders understand the IPD process, the alternatives available to them, and their responsibilities in an IPD project.

Why is owner alignment important?

IPD teams will do amazing things to achieve the owner's objectives, but they cannot do so if the owner does not make these objectives explicit. Owners must also ensure that their stakeholders are aligned around the IPD process and their responsibilities on an IPD project, to uncover any groups within the organization that resist change and any existing beliefs and processes that are inconsistent with IPD.

How do you align the owner organization?

Managed properly, the internal alignment process will create consensus among key owner groups. This can be a significant effort in larger organizations with many involved departments and stakeholders. The alignment process will involve selecting an appropriate IPD structural and business model from various alternatives, choosing an initial IPD contract model, and identifying real issues that must be considered when developing the IPD agreement. In most instances, the internal alignment process will lead directly to the creation of an owner/project–tailored IPD agreement that can be used to support the team selection process.

Determine the underlying business need or goal of a project. The owner entity must first address what it is trying to achieve with the project, knowing the internal owner stakeholders may differ in their specific approaches to achieving subgoals. What is the underlying business need or goal? Very few owners undertake a project just to get a building. For example, on a hospital project for a nonprofit organization, the goal was achieving better health outcomes for the local community. For a processing facility, the underlying goal was to improve the flow of products to reduce the cost of goods sold. The respective purposes affected how each project was designed and executed. Thus, you should determine your overall business need or goal on a project and ensure that this is shared and understood throughout your organization.

Expose, discuss, and resolve institutional impediments to IPD. It is not necessary to rebuild an organization to undertake IPD—many successful IPD projects launched despite hesitancy within the owner organization. However, do not ignore real institutional impediments: expose, discuss, and—to the extent practical—resolve them. Include legal, corporate compliance, risk management, or other groups that can affect the contracting process rather than hope that they will miraculously support a process they do not understand.

Develop a transparent and effective system for owner decision-making. Systems for owner decision-making should best fit the needs of each project. Some projects may succeed with a user-group-decision framework that identifies which committees and users make or influence certain types of decisions. Moreover, mapping which owner groups have information, influence, or approval authority will not only improve the owner's internal alignment and efficiency, it will allow the project team to understand and accommodate the owner's decision processes. Other projects can identify a single, accountable person who represents all users and is empowered to make binding decisions. Regardless of the system used, it must reliably deliver decisions that guide the project team and are supported by the owner's leadership. *(See Appendix 7 for an example of an unusual organization chart that maps a user-group-decision framework for a large public owner.)*

Hire an outside consultant. Most owners will require assistance in evaluating their options and working with their internal processes. The left-hand side of the Path to IPD Agreement diagram represents a process of internal reflections, usually in workshops of key owner personnel and stakeholders to clarify and define project values and goals. An outside consultant will bring skills, experience, and credibility to the process. In addition, many owners have found that an outside consultant helps with this process and improves communication among their various internal groups.

In summary, the key tasks for owner alignment are:

> Create shared understanding throughout the owner organization.

> Discuss and obtain commitment to IPD principles and processes.

> Develop a deep understanding of why the project is being undertaken.

A NOTE ON INSURANCE

It might seem premature to consider insurance before an IPD team has even been selected. If the project will rely on traditional design and construction insurance, then insurance issues may be deferred until the contract workshop. However, if more sophisticated insurance, such as an Owner Controlled Insurance Program with Project Specific Professional Liability/Rectification coverage is being considered, then promptly engage a skilled broker to begin assessing project requirements and market availability. Evaluating and obtaining insurance can be a lengthy process that interacts with the IPD agreement and can delay contract execution unless started early.

TEAM SELECTION
With Whom Do You Want to Work?

What is team selection?

Team selection is the process of selecting team members to deliver an IPD project. This includes determining which firms will be part of the risk/reward structure and which will not. Firms that are a part of the risk/reward structure are those that have placed their profit at risk and have the opportunity for increased profitability, based on the project's outcome. For the purposes of this guide, the firms participating in risk/reward are called the IPD team. There may be firms that are not part of the risk/reward structure. The totality of all the firms is referenced as the project team and individually as project participants.

Why is team selection important?

The goal of team selection is to involve project participants that are competent, engaged, and aligned to the project and the IPD process. Although every owner supports this goal, how team selection is achieved is highly variable.

How do I start IPD team selection?

Owners that have long-term relationships with members of the architect/engineer/contractor community may begin the process knowing some of the makeup of the IPD team. In that case, the owner and their trusted vendors will jointly develop a strategy for procuring the remaining members. This may be as informal as jointly interviewing parties that they already know, or it may be a formal request for qualifications/proposal (RFQ/RFP) process. *(See Appendix 1 for an example of an RFP in a market new to IPD.)* Owners with no pre-existing relationships, or who build infrequently, may prefer to issue an RFP for a complete IPD team, leaving to the participants the task of forming a cohesive team. Whether formal, informal, or something in between, plan the process and incorporate each team member. The owner is evaluating and selecting an IPD team, but the IPD team members are also evaluating and selecting each other.

The goal of team selection is to involve project participants that are competent, engaged, and aligned to the project and the IPD process.

TABLE 2: TYPICAL CATEGORIES FOR EVALUATING TEAM MEMBERS

CATEGORY	CHARACTERISTIC	EVALUATION	WEIGHT
Financial	Home Office Overhead Rate		
	Profit % (Similar Projects)		
	Rates for Team Members		
	Productivity Data		
Technical	General Competence		
	Specialized Capabilities		
	Digital Sophistication		
	Lean Capabilites		
	Prior Experience on Similar Projects		
Collaborative	Prior IPD Experience		
	Prior Experience with Other Team Members		
	Prior Demonstrated Collaboration		
	Learning Organization		

There is greater waste in bringing a party on too late than there is in bringing the party on too early. When in doubt, err on the side of engaging the IPD team members early in the project.

Clearly define financial terms used in the IPD contract and project. Terms such as *fee* may have very different meanings among IPD team members.

What else do I need to do to select the IPD team?

Determine who should be in the IPD team and when they should be added. The IPD team should generally be the principal designer, key consultants, the principal builder, and key trade contractors. Key in this context means parties that have a significant effect on project outcome. In some cases this list may become clearer over time. These parties need to closely collaborate with other parties or have a significant portion of the project cost within their scope of work. In some instances, if there are smaller scope trade contractors (such as low-voltage systems) that may drive a disproportionate share of change orders, consider including them in the IPD team to manage this risk. As a rule-of-thumb, represent no less than 50% of the project cost (design and construction) in the IPD team, although 75% or more is preferred.

The parties should be added when they can bring value to the project. In some cases, the entire IPD team is selected at inception. But often the IPD team is selected on a rolling basis, with some parties, such as the lead designer and lead builder, being selected and the remaining parties being jointly selected by the designer, builder, and owner with input from each party that has been selected. The important point is to ensure that there is joint involvement of the IPD team in selection of parties and that all parties are engaged sufficiently early to allow them to provide value to design development.

Select the IPD team based on best value. IPD teams are chosen based on competence and collaboration. However, it is important to have an understanding of finances at the time of selection. Explain the basic financial model in the RFP or similar document, including how the IPD team will be paid and how rates, overhead, and profit will be calculated. *(See Appendices 27 and 28 for templates to calculate direct cost, indirect cost,*

A NOTE ON SCHEDULING

The IPD team and other project participants have busy schedules. If you attempt to schedule key workshops after the team and other participants have been selected, you may incur significant delays due to the difficulty in scheduling multiple parties. Instead, publish the schedule for meetings in the RFP (or similar solicitation) to alert the responding parties that they are required to attend on specific dates. To do this, assemble your legal/consulting/insurance team (to the extent required) to coordinate their availability with the proposed schedule.

and overhead that can be used in the RFP/RFQ process.) Clearly define financial terms used in the IPD contract and project. Terms such as *fee* may have very different meanings among IPD team members. Defining terms will avoid surprises during negotiation and will allow fair basis for comparison for potential team members. Some owners use weighted factors and structured evaluations using tools such as Choosing by Advantages (CBA). Regardless of technique, it is important to look at IPD team selection holistically. While the dynamics may not be evident at the time of selection, keep in mind that the IPD team that can effectively merge their collective strengths will generally outperform a dysfunctional team of superstars.

The specific criteria for and weight given to IPD team interviews are project specific, but the general categories in Table 2 are relevant to most projects. These factors, or similar factors, can also be used to compare different alternative team members using CBA.

Interview the IPD team members, not the individual companies. The interview stage is a good opportunity to assess the IPD team's ability to work together. For this purpose, have the firms who are prospective IPD team members bring the actual parties who would be working on the project, rather than the marketing staff. They should be prepared to explain how they have worked together in the past and how they would work together for the current project. Some owners have asked prospective IPD teams to work on sample problems in real time to demonstrate their ability to work collaboratively. Research has shown implicit bias can limit equitable selection, therefore this is a good time to check in that you are using "bias interruption" techniques to ensure the selection process is inclusive. *(See Appendix 5 for an example of a partner evaluation matrix that is used to evaluate and select the IPD team.)*

Engage IPD team members as soon as they can reasonably contribute. In some projects, particularly those with a long development phase, procuring specific IPD team members may be deferred until they have contributions to make. However, remember that part of IPD's power is the ability to harness the specific knowledge of trade contractors/systems vendors *before* the design develops. There is greater waste in bringing a party on too late than there is in bringing the party on too early. When in doubt, err on the side of engaging the IPD team members early in the project.

Whether formal, informal, or something in between, plan the process and incorporate each team member. The owner is evaluating and selecting an IPD team, but the IPD team members are also evaluating and selecting each other.

---

3   Joan C. Williams et al., *Climate Control: Gender and Racial Bias in Engineering* (San Francisco: Center for Worklife Law & Society of Women Engineers, 2016).

THE CONTRACT WORKSHOP /
IPD TEAM ALIGNMENT

Does Everyone Know the Rules of Engagement?

_____

What are contract and team alignment workshops?

Contract and IPD team alignment workshops are some of the first opportunities for team collaboration. While they each have distinct goals, the contract workshop is focused on integrating the IPD team to achieve contract execution, while team alignment workshops focus on the team alignment process itself—how to establish a high-performing team.

The contract workshop is intertwined with the team alignment process. Separating them creates the distinct likelihood that the contract will not be consistent with the IPD team's values and goals and that the team will not understand the structure they must work in. Moreover, contract negotiation is often simpler and less contentious when those responsible for contract negotiation—legal, procurement, and risk-management personnel—have a solid understanding of IPD principles and can see for themselves how a team of disparate companies can work collaboratively on developing common goals and values. While we will discuss the contract and team alignment processes separately, we recommend that they be undertaken together.

Why are contract negotiation and team alignment workshops important?

The contract negotiation and team alignment workshops may be one of the first opportunities for the IPD team to work together collaboratively. Ensure they are well structured and facilitated by a person or persons who are well versed in IPD agreements and are skilled facilitators. If the IPD team uses external coaching, blend the contract and organizational sessions with meetings focused on communication skills, lean principles, or other learning that will be essential for project success. In addition, these events are opportunities for the team to begin developing interpersonal relationships. Many of the exercises should be conducted across disciplines and across companies to begin establishing a project-focused organization. Augment these exercises with informal social events, such as shared meals or activities.

The Contract Workshop

What is the contract workshop?

Regardless of the contract form being adopted, it is important to have a contract workshop. Properly speaking, it is a project kickoff and alignment session, as well as a contract negotiation. Ideally, the principal management personnel from each IPD team member, representatives of the stakeholders, and contract negotiators should participate. During multiday workshops, the number of participants may decline as the focus moves from business models and project organization to contractual specifics. The final session may be solely for contract negotiators and key decision-makers.

Why is the contract workshop important?

The contract workshop has multiple objectives. Because some or all of the participants may not have IPD experience, they will tend to make assumptions that are incorrect. The workshop is a good venue to address questions and concerns regarding IPD and the project. The workshop is also an opportunity to educate the IPD team regarding IPD principles, their roles in an IPD project, and how the IPD agreement will function. Have the workshop occur before substantive discussion of the contract to assure a common understanding of IPD and to provide the correct context for discussion and negotiation. IPD projects require honesty, transparency, and an ability to view issues from multiple perspectives. The workshop is an opportunity for the owner to model these values and set the project's tone.

Who should be involved in the contract workshop?

The owner, IPD team, and other stakeholders/users are included in the early sessions to ensure that the underlying reasons for the project are properly expressed and considered during the negotiation process.

How do I run a contract workshop?

The contract workshop should be facilitated to assure that all participants take part and that no one leaves with unanswered questions. In addition, the facilitator should document the key decisions made and develop a schedule and assignment of tasks for contract completion. Most contract workshops focus on the following tasks.

Training on IPD principles. Not everyone will have come to the workshop with the same knowledge and experience with IPD. Given social pressures, the less experienced will likely ask few questions and will appear to agree, although they may not be committed. A brief IPD refresher before discussing the business model and contract terms will establish a level of understanding shared by all IPD team members, including the owner.

Jointly create and document goals and values. The contract workshop is an opportunity to align the entire IPD team to the project's core values and goals and to begin jointly developing structures and procedures for managing the project. Working together to create and document goals and values is a step toward team commitment. Moreover, in IPD the project management team (PMT) is responsible for achieving the project objectives; therefore, explore project objectives, values, and goals during the workshop before focusing on contract terms.

Jointly make contract decisions. There are many issues and decisions that need to be made by the entire project team. These range from the basic structure of the IPD agreement (e.g., a multiparty agreement or poly-party agreement), governance structure, cash flow and financials, insurance, extent of liability waivers and limitations, and other concerns. It is important that the IPD team understands how all of these items actually work in practice. The IPD team needs to be able to describe how it wants to work in order for the contract to document these decisions.

Test the contract in multiple scenarios. For many people, reading contract language does not provide them with a clear understanding of how the contract will operate. A better approach is to dedicate a portion of the workshop to "testing" the contract against possible scenarios. IPD team members should ask,

"What would happen if..." Then, with the aid of the facilitator, explore how the contract would handle the situation. This can provide deeper understanding than merely reviewing contract language.

Finalize the process to contract execution. The last phase of the contract workshop focuses on the tasks required to complete the IPD agreement and establishes a schedule for the creation of required exhibits, the process for reviewing the project-specific-contract draft, and a date for contract execution.

Team Alignment Workshop

What is the team alignment workshop?

The IPD team alignment workshop is the first formal team event in a continually evolving process (duration of which varies widely) that spans from conception to completion. The workshop helps to begin creating a project culture and to commit the IPD team to project values and goals. The resulting agreements are often embodied in a project charter, mission statement, or Conditions of Satisfaction (CoS). *(See Appendices 2 and 4 for examples.)*

Why is the team alignment workshop important?

IPD takes multiple firms and molds them into a virtual organization aligned to achieving project goals. Therefore, IPD team alignment has two primary functions: it creates clarity, consensus, and commitment to project goals, and it begins to shape a project culture that transforms a group of individuals into a high-performance team. Both of these elements continue throughout the project. On-boarding, discussed more fully in Team Maintenance, is an extension of the work completed in the team alignment workshop.

IPD team alignment is also a first stage in enhancing communication and building trust among IPD team members. A strong team has a strong team culture. IPD team members lead by example, trust one another, and are willing to listen to other's ideas. Strong team

alignment means having a team that is trained in collaborative IPD behaviors, has aligned their professional goals with project goals, and has well-established trust. Structure the exercises to teach communication and leadership skills to the team. Use this workshop to also build relationships among the team members by working together in a collaborative, controlled environment. *(See On-boarding for more on how to continue these efforts when adding new team members.)*

How do you run an IPD team alignment workshop?

Hire a facilitator to lead the workshop. There is a natural tendency for IPD team members to defer to the owner and for trade contractors to defer to a construction manager/general contractor who they depend on for other projects. This can reinforce traditional roles and work against collaborative IPD culture. A facilitator can neutralize some of these tendencies by structuring active exercises of small multidisciplinary groups and can monitor and guide the process to ensure participation and interaction.

Each facilitator has her or his preferred techniques and tools. In most instances, a combination of stakeholders, owner's personnel, and IPD team members will be arranged into small groups to jointly discuss and create a list of key values. These will be presented to the entire group using stickies, cards, or easel pads. The role of the facilitator is to ensure that the views of all parties are expressed, that no "strong voices" dominate, and that all parties have or develop a very similar understanding of the goals. The underlying reasons for the project should be clearly identified. So, too, should the legitimate goals of the IPD team members. The owner cannot expect a team to be committed to their values if the owner has no consideration of the team's values.

Use the workshop as the basis for a project charter or CoS.[4] The workshop product becomes the basis for a project charter (if one is used) or becomes the value list (CoS) that the PMT uses to assist decision-making.

---

4 For more information on how to get to Conditions of Satisfaction, see M. Fischer, H. Ashcraft, D. Read, and A. Khanzode, "Managing with Metrics" (chapter 11) and "Collaborating in an Integrated Project" (chapter 13, particularly section 13.4) in *Integrating Project Delivery* (Hoboken, NJ: Wiley, 2017).

These documents are distinct from the project goals and values, which state and clarify the desired outcome of the project and determine what is believed to be the most important ideals of the project. The CoS determines the expectations or requirements that need to be satisfied to deem an outcome as being successful. Whereas a project charter determines roles, responsibilities, and team behavioral goals, setting expectations around behaviors and decision-making.

Often, the workshop document is signed by the participants and publicly displayed in the project co-location space or another prominent location. It may also become the foundational material for a project mission statement. In addition, the goals may be tied into the contract risk/reward program or may otherwise influence the contract terms. Ensure that the product from team alignment is a constant influence and guide for the project.

Provide social opportunities for participants. The IPD team alignment/contract workshop is an opportunity for the participants to engage socially and professionally. Host a few organized events—they need not be expensive or elaborate—to create opportunities to build personal bonds that will build trust. Pizza and a beer, or a low-impact game, are better than a formal dinner.

In summary, the key goals of the IPD team alignment workshop are to:

> Create a common and accurate understanding of the owner's goals.

> Develop clarity regarding the goals and commitment to achieving them.

> Expose concerns IPD team members may have regarding people, the process, and the project.

> Include IPD team goals in the project goals.

> Obtain owner commitment to all the goals.

> Develop or refine CoS.

> Develop or strengthen interpersonal relationships on the IPD team.

> Model and teach collaborative behaviors and techniques to team members.

A good IPD team alignment workshop should set the stage for project execution. But in any project, additional individuals and parties will be added after the IPD team alignment workshop and must be integrated into the project and aligned with its goals through an effective onboarding process.

Related chapters include: **On-boarding** (pg. 93)

A NOTE ON CONTRACT FORMS

Currently, there are a variety of form contracts available. The principal North American association options are the AIA C-191, ConsensusDocs 300, and CCDC 30. These can be good starting points, but all will require completion and modification before use on a specific project. In addition, there are proprietary forms that have been used widely, particularly those based on the early Sutter Health agreements and the Hanson Bridgett LLP forms. One advantage of the proprietary forms is that they continuously embed lessons learned from real projects.

If you are unacquainted with the various agreements, seek assistance from a construction attorney who has handled multiple IPD projects and can help you assess the best approach for the project. We strongly discourage creating an entirely new contract from scratch. It is very expensive and, unless the drafter has considerable IPD experience, is unlikely to be as thorough or effective as existing agreements.

What Do I Do after the Workshops?

Assign leaders to each workshop task. There will undoubtedly be tasks assigned during the contract workshop and there will be contract exhibits or other information that must be generated and vetted by the team. At the workshop, develop a schedule and assign each exhibit and task to a leader who will be responsible for selecting and leading a subteam that will complete the task or exhibit.

Prepare a final contract draft. Prepare a final contract draft and circulate it among the parties for final review and comment. In a poly-party setting, it can be confusing if each party physically marks up a copy of the contract with strikethroughs, additions, and comments. Instead, collect and manage the comments in a matrix format, and resolve the remaining technical issues by web-conference or in person.

Develop commercial terms in parallel with the legal terms. Develop commercial terms (e.g., anticipated profit, risk/reward plan, overhead reimbursement) in parallel with the legal terms. The commercial terms need to be very clear and completely transparent. On larger projects, there is often an early financial audit to determine the appropriate calculation of reimbursable rates, overhead, and profit.

Conduct a contract-signing event. Technically, there is no need for a signing event. However, many IPD teams find that having a celebratory signing reinforces their commitment to a collaborative project.

A good IPD team alignment workshop should set the stage for project execution.

VALUES, GOALS AND CONDITIONS OF SATISFACTION

These terms are often used interchangeably, but they express different levels of granularity. Values are fundamental, high level beliefs. Environmental stewardship, or respect for people, are values. Goals reflect actions that are consistent with the values. Reducing greenhouse gasses or creating trust are goals. Conditions of Satisfaction (CoS) reflect specific commitments. Achieving a net zero energy project or having a planned percent complete of 82%, are CoS. Whenever possible, CoS are measured for management and/or compensation purposes.

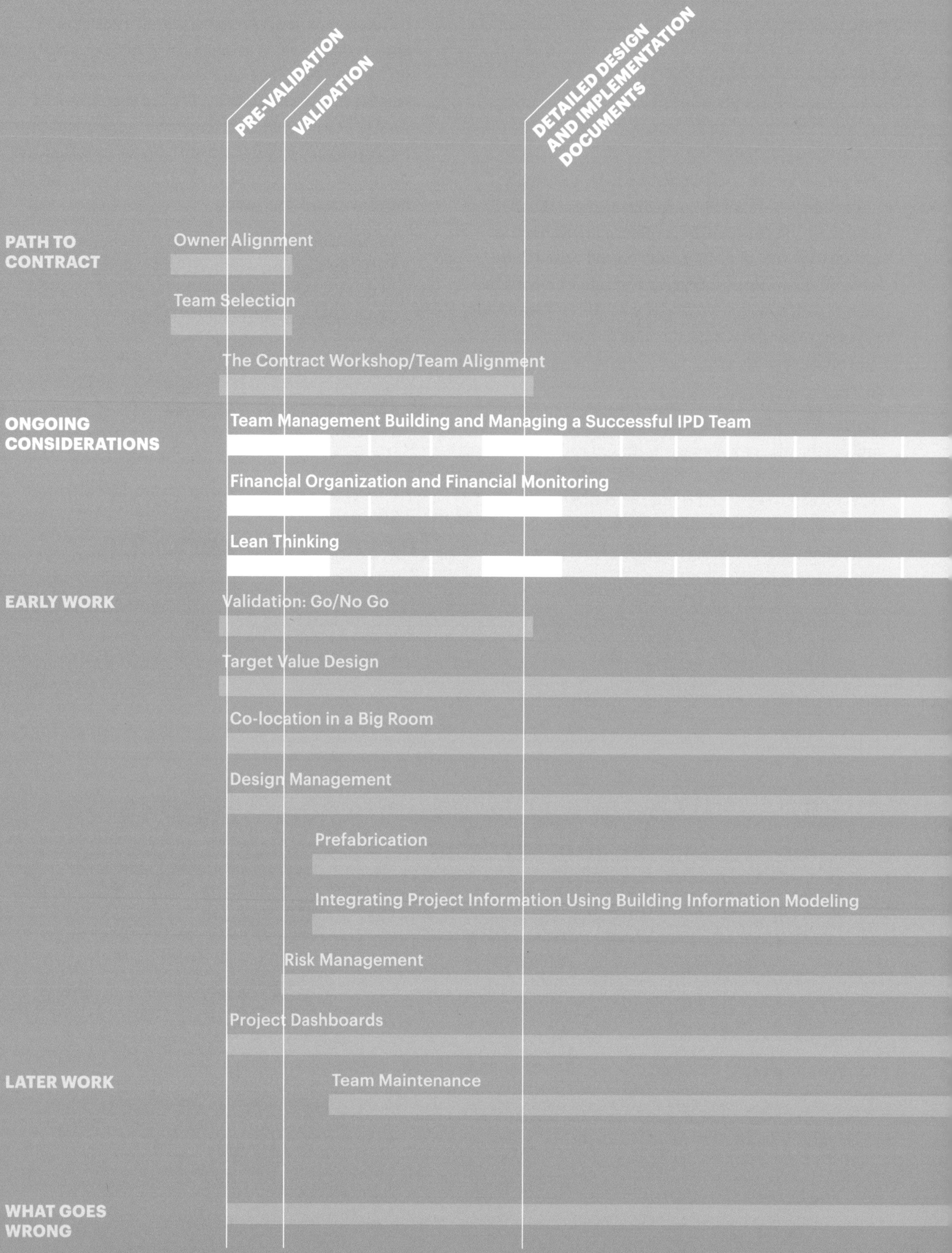
PRE-VALIDATION
VALIDATION
DETAILED DESIGN AND IMPLEMENTATION DOCUMENTS
PATH TO CONTRACT
Owner Alignment
Team Selection
The Contract Workshop/Team Alignment
ONGOING CONSIDERATIONS
Team Management Building and Managing a Successful IPD Team
Financial Organization and Financial Monitoring
Lean Thinking
EARLY WORK
Validation: Go/No Go
Target Value Design
Co-location in a Big Room
Design Management
Prefabrication
Integrating Project Information Using Building Information Modeling
Risk Management
Project Dashboards
LATER WORK
Team Maintenance
WHAT GOES WRONG

Closeout

FROM BEGINNING TO END
Ongoing Considerations

As an IPD leader, you will need to spend a significant amount of time at the beginning of the project laying the groundwork that establishes team and project management, financial organization and processes of maintenance, and efficient management processes using lean tools. Once this groundwork is set, maintain the team culture, project finances, and management processes throughout the project.

The following section covers the ongoing concerns that an IPD leader must consider from beginning to end. Team Management covers how to build a strong team culture that provides psychological safety and feelings of mutual trust for team members. This is key for ensuring team collaborative behaviors and promoting transparent communication around team and project needs from design to construction. Team management also includes how to set up a team organizational structure that identifies leaders and decision-makers, as well as processes for decision-making.

The Financial Organization and Financial Monitoring chapter covers how to set up an open-book financial system that will help drive success on the project. This section also discusses how to monitor finances throughout the project, as well as how to obtain a detailed cost estimate with appropriate trade contractor input.

The final chapter in this section is on lean thinking: the principles and processes that will improve efficiency on your team. IPD provides an optimum environment for lean to thrive. Establishing these principles and processes early in the project and maintaining them throughout will help ensure that your team will achieve success.

TEAM MANAGEMENT
How Do We Build and Manage a Successful IPD Team?

Why is team management important?

IPD project teams have earned a reputation as high functioning and resilient. When IPD project teams are stressed or challenged, they are able to pull together as a team to resolve the situation, make adjustments as needed, and ready themselves quickly for the next challenge. IPD project teams have strong collaborative cultures marked by mutual trust, respect, accountability, transparency, and psychological safety. These markers extend to most or all project participants, regardless if they are in the group of IPD team members whose companies have put their profit at risk. IPD project teams also rely on clear communication, which leads to positive outcomes, such as reduced RFIs. To a certain extent, for the IPD team, these characteristics emerge because of the intensive IPD project-contract discussions and other early interactions. However, team building is strengthened by the intentional actions covered in this section, which can also extend the positive IPD team culture to include the broader project team. Maintaining the team over the course of the project is covered in Later Work.

What are the characteristics of strong team culture?

Among the markers, psychological safety and the related mutual trust and respect are key for innovation. IPD project teams who take risks and know how to manage their interdisciplinary expertise will reach levels of creativity that nonintegrated teams cannot.

What are the expectations for members of the IPD team?

Project managers need to expect a large amount of time needed for relationship building and planning. Project managers need to develop direct relationships with all team members, including trade contractors. They also need to participate in project implementation team (PIT), project management team (PMT), and senior management team (SMT) meetings, as well as design meetings. This requires more planning and support time than in a traditional project. Plan and estimate for this time commitment.

Team members need to make the time to come to the table prepared. Team members need to commit to the time required to prepare for project questions in meetings. They need to bring information to the table that will help the team make decisions.

Trade contractors may have difficulties making the time commitments needed on IPD projects. Trade contractors are traditionally not set up for an IPD organizational structure. IPD projects require contractors to staff differently than projects consisting of lump sum work. Pay attention to trade contractors that are not at risk. Manage, treat, and work into the project your non-risk trade contractors just as you would with your trade partners. Ensure that they also understand the IPD process and culture and candidly discuss the costs incurred for time required.

What goes into managing the team and who leads this process?

There are three main aspects to managing an IPD project team: establishing/maintaining team culture, developing the structure of the team, and managing the work of the team. For some teams, the same person is the lead for all three; for others, different people lead different aspects or fill roles that can shift over time, depending on the phase of the project. There are good reasons for why the owner, the architect, or the general contractor can play lead roles. Finding the right fit depends on your team's needs.

How Do I Establish Team Culture?

Team culture needs to be established and maintained over the duration of the project. *(See Appendix 18 for an innovative method of establishing team culture.)* Many of the early work discussions on values, contracts, and roles serve a dual purpose: they help the team make decisions and lay the foundation for strong team culture. Is it important to distinguish between a member of the IPD team (whose profits are in the risk-reward pool) and other project team members (participants engaging in the project with conventional contractual agreements)? All team members contribute to the culture of the project team and should be expected to support the goals and values established by the IPD team. The IPD team members are particularly responsible for modeling positive behavior and should be expected to fully understand IPD principles and how they work. Teams are not static: the best project teams will maintain a strong team culture even as they evolve over the course of project phases, personnel changes, and challenges. The following are ways to establish and maintain team culture.

Use on-boarding to set behavior expectations and ground rules for the team, and establish an off-boarding process well before any issues arise. Establishing team culture means establishing expectations and rules around behaviors, team values, as well as communication and management processes. This also includes setting expectations and rules around how and why team members would be removed from a project. *(See On-boarding and Off-boarding for more on setting expectations and rules for the team.)*

Measure and track goals related to team performance. Measure team performance through team health checks, reliability, and other metrics. *(See Project Dashboards for more on metrics.)* If your metrics indicate that the team is not achieving these goals, have a candid conversation with the team to determine the cause of

ELIMINATE THE FIGHT OR FLIGHT INSTINCT AND MAKE IT POSSIBLE TO BE CHALLENGED

When challenged on an idea, there is a desire to defend one's idea at any cost, which can halt productive discussion. One contractor suggested that you can intentionally shift out of having a fight or flight reaction by coming in with the right attitude. You can say to yourself, wait a minute, as stupid as that sounds, he just might be right.

TYING PROFIT FOR THE COACH TO THE PROFIT POOL

While many coaches are not tied to the outcomes of a project, in one owner's case, the team wanted to tie the coach's profit to project outcomes. The owner reported that this worked because "the team now felt that the coach had skin in the game with them. So he's looking out for their best interest, not just showing up to build the team for a day rate."

the problem and develop countermeasures together to get the team back on track. *(See Appendices 19, 20, and 21 for examples illustrating how teams assess and track their performance on the project.)*

Co-locate the team. Co-location can be a powerful, positive force if the team culture is strong. Conversely, co-location can be a source of tension if the team has not established a positive culture or if the expectations of roles and engagement are not clear. For complex projects, co-location is typically an essential investment. *(See Co-location in a Big Room.)*

Optional processes that can be helpful:

Conduct regular team check-ins. There are two types of regular check-ins: one focused on team culture and another on project information flow. For larger projects, these should be done daily.

Ensure project information flows. Use regular communication, such as a huddle or scrum for project updates. Ensure that communication flows amongst the team as a whole, and the PMT. These communications can occur as a morning huddle or scrum between the team and team leaders. These conversations should involve asking, "What did you do yesterday to advance the goals of the project?" "What are you doing today to advance the goals of the project?" "And what stands in your way?" For smaller projects, weekly communication may be sufficient.

Check in on resource allocation between disciplines. Sometimes team members do more work than needed or do redundant work that another team member is doing. This can waste project time and budget, this is particularly true for IPD team members who are paid for their time and overhead. Check in on team-resource allocation to see what they are doing versus what they should be doing. *(See Appendices 23 and 24 for tracking IPD team time and costs.)* Emphasize the need to work efficiently and find ways to improve efficiencies during design and construction.

Create a learning environment. Continuous learning opportunities can be a regular part of work for the team covering topics such as: technical issues, processes, leadership, and personnel/personality.

What are the benefits of using a coach or facilitator on the team?

Having a team coach or facilitator can help the team learn how to solve problems creatively and share in risk/reward. Coaches can come in multiple times a

COVERING THE COSTS OF A COACH

One owner recalled an early IPD project during which the team indicated that they needed help but felt that the expense for the coach should fall to the owner. The owner responded to the team saying, "I am going to front the cost and we are going to teach the team. If the training works, your savings covers the cost. If the training does not work, I cover the cost." At the end of the project, the team covered the cost of the consultant. The owner added, "It was not a matter of whose responsibility it was to pay, it is about solving the problem through a fair and equitable approach."

Another team, in recognition that the coaching and training would benefit their companies long after the end of the project, elected to divide the cost of coaching equally among signatories and pay for it outside of the project.

HOW ARE IPD TEAMS STRUCTURED?

As with any construction project, IPD projects involve dozens of companies and hundreds of people. Companies in the IPD Team are part of the risk/reward structure; other companies are referred to as non-signatory. The management structure of a typical IPD project is fairly straightforward, including the three groups, and their associated responsibilities, as shown below.

The Project Management Team (PMT) and the Project Implementation Teams (PITs) are where the real work gets done.

PROJECT MANAGEMENT TEAM (PMT)

The PMT is the project's administrative workhorse, making the tough decisions and monitoring financials. The PMT always includes the owner, architect, and contractor. In some cases, other members of the IPD team may serve on the PMT.

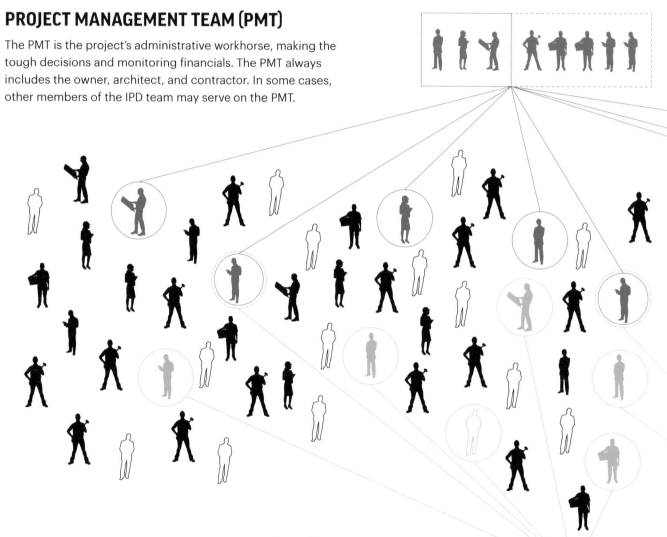

PROJECT IMPLEMENTATION TEAMS (PITs)

Made up of diverse stakeholders organized by areas, PITs drive innovation and value into (and waste out of) the project. PITs can include all members of the team—the PMT, signatories, non-signatories, owners, architects, contractors, trades, suppliers, resulting in small multidisciplinary teams. Common PITs include structure, mechanical, electrical, envelope. The specific number of types of PITs will be determined by the team.

PIT 1

EXAMPLE PIT CONFIGURATIONS

SENIOR MANAGEMENT TEAM (SMT)

The SMT always handles dispute resolution and backup, as required. Often, they also conduct contract negotiations and resolve questions of scope change, but this can alternatively be done by the PMT. The SMT is composed of one C-level executive from every party that signs the IPD agreement.

EXAMPLE SMT FOR SIX PARTY AGREEMENT

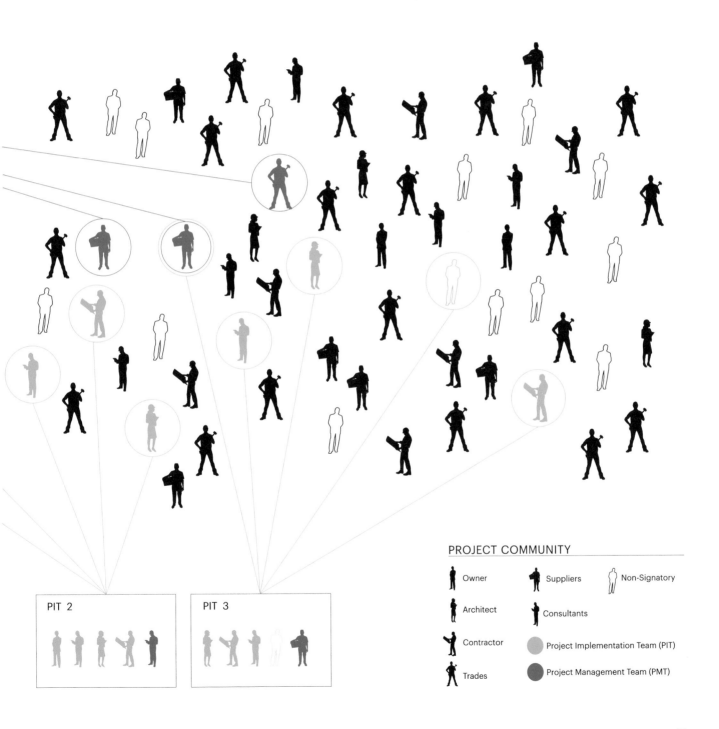

PIT 2

PIT 3

PROJECT COMMUNITY

Owner   Suppliers   Non-Signatory

Architect   Consultants

Contractor   Project Implementation Team (PIT)

Trades   Project Management Team (PMT)

month to once every few months and can help with on-boarding and training the team in facilitation skills. Generally, frequent coaching is most effective in the early phases focused on the IPD team. The fact that the coach is not as attached to the outcomes of a project and is solely focused on how to get the team to be successful can be a benefit. There are many options for who covers the cost of the coach: sometimes it's the owner; other times the cost is shared within the team.

Focus the facilitator on training the team to function on its own. After the facilitator leaves, you want the team to remain collaborative and high performing. Make sure that the IPD team is still seen as the leaders in the room. *(See Team Maintenance.)*

How Do I Establish Team Structure?

To establish a strong team culture, you need to establish a clear team organizational structure that identifies key leaders and decision-makers. Team structure should also include a clear scope of work for IPD team members as well as other project team members. Clearly the IPD team is deeply engaged with steering the project finances, CoS, and team culture. Within the IPD team, Hanson Bridgett and other contracts name a PMT and SMT with specific roles, interactions, and scenarios for when the PMT group passes decisions on to the SMT. The contract may also identify PITs, which include subject-matter experts and key users. Regardless of the language, project teams need to identify the level of involvement of key stakeholders and determine whether they influence the project or have authority in project decision-making. Identify whether any of these key stakeholders will be members of the IPD team. Some teams find a diagram of the

DO NOT CONFUSE LEADERSHIP WITH AUTHORITY

Owners and other team leaders need to respect the knowledge, skills, and expertise of other team members. Just because you have authority as a high-ranking decision-maker on the project, does not mean that other team members are not leaders and experts in their own right, with their own innovative ideas for a project. Listen to the ideas of others, including those coming from the trade contractors that may not have a seat at the table.

WHAT ELSE SHOULD I CONSIDER WITH TOOLS AND PROCESSES?

When managing and planning tools and processes on a project, also consider the following:

> When things go wrong, be hard on the process, not on the people. Look to use the people and team to quickly assess the problem, its root cause, and arrive at process improvements.

> Pool and share equipment and software. Sharing equipment and software can help manage project costs and improve and standardize tracking.

> Ensure that initiatives and processes are sustainable across the team. Often, teams set up "extracurricular" processes, additional work outside of project work, but they are not sustainable in the long run. Go slow with extracurricular processes and ensure they are sustainable across the team in the long term.

structure in an organizational chart is helpful, while others have found decision matrices to be helpful. *(See Appendix 7 for an organizational chart example and Appendix 9 for an example of a decision matrix.)*

Define and use PITs. Organize the project team into small, nimble, multidisciplinary PITs to conduct deep dives on specific project needs (e.g., mechanical, envelope, and structure are common, but also non-design topics such as innovation or lean). This helps project team members take ownership of a certain part of the project and budget. For example, architects, mechanical engineers, and structural engineers can work as a PIT on building systems. Assign PITs an area within the Big Room and a time during which they report out on their work to the team. Provide PITs the authority to bring people into their group at the right time. Some PIT members will be in the IPD team and may also play roles in the PMT while other PIT members are not even on the IPD team.

Establish PIT leaders. Each PIT should have a PIT leader. The PMT selects the PIT leaders, or the role can develop organically, filled by a PIT member who has the capacity, interest, and expertise for a leadership role.

Define a decision-making process or system. You need to have the right people in the right meetings for decision-making. Ask the following questions: Who on the team must be in the room to make decisions, What decisions can they make? Use a system to document the key decision-makers for specific types of decisions. Or develop a process by which the right people are determined for each case. Remain flexible enough during the project to be able to balance between who needs to be involved and who else on the team should be added as decision-makers at certain points in the project.

DIFFERENT ROLES, DIFFERENT NEEDS, AND DIFFERENT IDEAS OF SUCCESS

Contractors and designers need different things to be successful. For example, in pull planning or one-piece-flow, designers should ask, "What do I need from the contractors to be successful?" If you are a contractor, remember, designers think they know what you need, but they may not.

RETHINKING RESOURCE ALLOCATION

One owner described the importance of thinking about the team's time as a total number of hours to generate the most value. Since the most qualified person will need to do the most specialized work, a trade contractor may be asked to do a drawing to free up the designer's time for tasks that require their specific expertise. The conventional way of thinking about billable hours tied to tasks is, as the designer describes, "If I am not drawing that, you are taking my money away." The owner's response from the IPD point of view shifts away from taking away billable hours to allocating the designer's time to a better use: "I want to pay you to go out and be creative and design some stuff because that what you love to do; it's what I need you to do."

How Do I Plan for Decision-making and Generally Managing the Work of the Team?

Because IPD project teams do their most intensive work early in the project, they do not have the luxury of organically developing their decision-making processes. Intentional investment to create effective processes and structures is important. One of the first things the IPD team needs to do is assign work and decision-making responsibilities that cover the duration of the project. For example, in earlier phases of the project, the designer will often take the lead, whereas when it transitions to construction, the general contractor or a trade contractor may lead the effort. Documentation of these responsibilities is sometimes called a task or responsibility matrix. *(See Appendix 6 for an example.)* Documents like these do not inhibit collaboration but gives team members ownership over their scope of work and provides clarity around who is leading specific work efforts. This helps the whole team know who is responsible for what, who needs to be involved with making decisions, and what decisions have been made.

Determine roles and responsibilities early. The project team needs to know their roles and their responsibilities on the project. Bring your PMT leaders together during design/predesign to determine roles and responsibilities.

Define deliverables and time needed to complete deliverables. The project team needs to know what deliverables are expected and when they are due. As a team, define the deliverables needed and when they are required. *(See Design Management for more information on how to budget time for these activities.)*

Determine what information is required from team members. Team members should identify what information they need to accomplish their deliverable and who is responsible for this information. Have the team identify the information source, amount of information needed, degree of accuracy, and the time needed to produce this information. *(See Integrating Project Information for Building Information Modeling for more information on how BIM deliverables are defined.)*

Integrate information with the delivery schedule. After the identification of expectations, deliverables, the information required, and time needed to complete deliverables, the information should be integrated into the delivery schedule. Specifically, owner decision-making needs to be predictable and well planned.

Ensure face-to-face team time each month. If you are not co-locating your team, make sure that team members have face-to-face team interaction on a weekly or monthly basis. The regularity of team meetings depends on the project phase. For example, during design, a biweekly meeting may be sufficient for setting project direction, while during construction, project teams may meet more frequently.

Determine how to document decision-making. Decide how the team should document the decision-making process to fulfill the team's needs. Any type of documentation should help plan decisions and prevent the team and owner from revisiting decisions once they have been made. The majority of IPD teams rely on A3s during the construction process for options analysis, agreements, and sign-offs. *(See Appendix 12 for an example of an A3.)* Another option is to use detailed meeting minutes or logs.

Related chapters include: **On-boarding** (pg. 93), **Off-boarding** (pg. 99), **Project Dashboards** (pg. 83), **Co-Location in a Big Room** (pg. 71), **Team Maintenance** (pg. 93), **Design Management** (pg. 73), **Integrating Project Information Using Building Information Modeling** (pg. 79)

FINANCIAL ORGANIZATION AND FINANCIAL MONITORING

How Do We Manage the Money from Contract to Completion?

What is financial organization?

One of the most powerful components of the IPD process is the financial organization of major design and construction partners through an IPD agreement with shared risk/reward. Different companies represented on the IPD team will either succeed or fail together within the IPD-contracting structure. Therefore, the team will be highly motivated to work together to optimize the whole project instead of their respective pieces.

Design and construction teams must negotiate business terms in a transparent fashion. This includes providing and collecting information from IPD team members on overhead and profit percentages, labor rates, contingencies, allowances, at-risk amounts, risk/reward splits, profit release, and clawback provisions, among other financial information.

Why is financial organization important?

For the shared risk/reward mechanisms in the contract to drive behavior, profit must be separated out from cost and overhead. It will not work if one party has 100% of their profit at risk in the agreement and another member has less than 100%. This type of setup would mean that second member is guaranteed some portion of profit on the project regardless of the outcome.

It is important to set up a system in which nobody wins if other parties lose. This means that one's success depends upon the success of all. This creates a system of mutual support in which higher-performing team members will raise the performance of lower performers instead of mimicking their poor behavior or trying to isolate themselves from the poor-performing team member.

How do you set up a financial system?

There are multiple components to setting up an open-book financial system. These components include—but are not limited to—profit, overhead, and labor rates. Keep in mind that designers and contractors typically develop costs in different ways. *(See Appendices 27 and 28 for examples of cost calculators used to establish billable rates of IPD team members. See the table describing these differences on the next page.)*

Auditing. The primary goal of audits is to ensure that there is no profit hidden in the overhead or labor rates. A secondary goal is to ensure that the IPD team members can recover their costs but not earn income from two different sources. For example, if a cost is included in the overhead, it cannot also be in the cost of work or labor rates. Likewise, if it is included in the labor rate (e.g., a cell phone or gas), then it cannot be included in the cost of work or overhead calculation.

	CONSTRUCTION TEAMS	DESIGN TEAMS
Profit	Profit is typically the one component of business terms that is not negotiated or audited. IPD team members will propose a profit during the selection process (usually a percentage of cost), and, if selected, that is their profit percentage. When the contract value is negotiated, their percentage turns into a lump sum amount based on their specific contract value at that time. The percentage is then put 100% at risk.	The amount is generally estimated as a percentage of the overall total billing rate. The amount is proposed by the firm during the selection process or negotiated by each IPD team member after the selection process is complete.
Overhead	Overhead is typically a proposed percentage by each IPD team member during the selection process. Once the IPD team is selected, an audit can be performed to confirm that the overhead percentage is supported by the financial history of the business. The rate may be adjusted after the audit. Overhead is applied as a percentage on cost of work, or billed to a max, or set as a lump sum for the project, depending on project specifics.	Design teams typically recover overhead through a multiplier on their direct labor cost. This can be proposed by the firm and confirmed by an auditing process.
Labor Rates	Labor rates are typically proposed during the selection process based on a previously supplied labor-rate build. The rates may be audited and adjusted if they do not match the financial history of the firm.	The design team typically recovers direct labor expenses, indirect labor expenses, indirect labor expenses, and overhead through their labor rates. An IPD team will typically make sure that the actual rates charged match what the employees are paid and that the burdens are justified by company financials.

What is the Overall Contract Value?

The overall contract value for an IPD project is a combination of all design and construction costs, including the total of all fixed profit amounts for IPD team members plus a negotiated contingency for the team. Typically, there is a shared savings plan negotiated by the IPD team for finishing the project under the negotiated contract value. This is set at some point during the design process, as early as immediately after validation or as late as the middle of construction. There is more opportunity to drive down costs if the shared savings plan is set earlier in the design process, since there is more certainty around the costs as the project progresses. Have the overall contingency and savings plan reflect when the IPD team commits to the contract value.

A clear system for tracking financial performance—aggregating all known information and presenting it in a clear format—is critical for project success.

Other things to consider:

Align risk and reward. Risk, profit, and contingency need to be balanced for each IPD team member. Negotiate together the amount of profit at risk, the amount of contingency that protects profit at risk, and the amount of additional profit that can be earned through shared savings; and ensure they complement one another. A fair deal typically does not have the IPD team member putting 100% of their profit at risk with the ability to only make 10% additional profit if the project finishes significantly under budget. Typical ranges of enhanced profit for projects finishing significantly under budget fall between 50% and 100% additional profit.

Consider overhead caps. Consider whether the owner or the IPD team wants to put a cap on overhead reimbursement for IPD team members. If an IPD team member goes significantly over their estimated budget, the team may not want to be in a position of providing additional home-office overhead compensation as well.

For the trade partners and the construction manager/ general contractor (CM/GC) this can be done by capping their overhead reimbursement (e.g., 5% added to a cost up to $200,000, after which the cost is reimbursed but without additional overhead). Another way to handle this is to make the overhead for the trade partners and CM/GC a lump sum amount similar to the profit but guaranteed. In this case, the IPD team member keeps savings if they finish under budget and do not recoup additional overhead if they spend more than their estimated cost.

Design IPD team members can cap overhead either by using a different labor-rate multiplier when they pass a certain cost threshold or by agreeing to change the rate by a preagreed amount when they pass that same threshold. This is something for teams to consider.

Plan ahead for warranties. Warranties can create confusion at the end of a project. The larger the project, the more important it is for an IPD team to have a clear plan around how warranties are to be handled at the end, primarily for any one of the IPD team members. Because they have a guaranteed cost contract, much of the early warranty work may be done on a cost-reimbursed basis and raise the overall cost of the project.

Moreover, different IPD team members budget and account for warrantee costs in different ways. Some team members may bill a small portion each month that is held for warrantee work. Others may pay warrantee costs out of their overhead allocation (and have it factored into the overhead rate in the audit). Have the IPD team discuss early in the development of the budget how warrantee work will be tracked and paid for on the project.

It is important to draw a firm line at the end of a project when the cost-reimbursed nature ends. Any costs for warrantee work after that period will be handled according to the warrantee plan.

Audit as early as possible. If an audit will be performed on a project, do it as early as possible. This minimizes the potential for misunderstandings if terms are modified after incurring significant costs.

What is financial monitoring?

Financial monitoring is a comprehensive financial-forecasting system in which the team and owner closely monitor the performance of each team member and—in turn—the budget of the overall project. IPD teams should meet at least monthly to review the projections of each firm and the aggregate projection. Each IPD team member's firm produces their own financial-projection worksheet to show their contract schedule of values, the cost to date, and the expected future cost for each of the line items in the schedule of values. *(See Appendices 23 and 24.)* The net monthly change for each firm is aggregated, usually by the general contractor, with the net risk register items to produce a summary sheet showing the current financial health of the project. This is usually shown on a graph with a trend line and compared to the contingency. *(See Appendix 25.)* Constant financial monitoring will allow the IPD team to discuss issues affecting cost and make changes if possible.

Why is financial monitoring important?

An IPD team needs to understand the current financial status of a project, where the project is likely to finish, and how the trend compares to the contract terms. Therefore, a clear system for tracking financial performance—aggregating all known information and presenting it in a clear format—is critical for success.

Unlike traditional agreements, IPD contracts tie the financial performance of individual design and construction firms to the overall outcome of the project. If a design firm burns through more hours than planned, this will impact the profit of all design and construction firms participating in the risk/reward program. Since costs are guaranteed to firms in the program, any one firm spending more time than budgeted will impact the financial outcome of the entire project. A firm may exceed their budgeted hours because they underestimated the scope of work or installed work incorrectly. In a lump sum contract, if that firm erred in their planning, they bear the responsibility for costs of the additional time to correct their work. But in this agreement, all IPD team member costs are guaranteed—if additional time is needed, profit is reduced for everyone.

How is financial monitoring done?

Financial monitoring begins with gaining certainty in your numbers. This means you need to take the time to do the validation step well. *(See Validation for more information.)* It also helps to produce a contract document with:

> Confirmed goals and logic of design;

> Scope;

> Budget;

> Assumptions; and

> Contingency or risk register.

How Do I Get a Detailed Cost Estimate with Appropriate Trade Partner Inputs?

To get a detailed cost estimate with appropriate IPD team member inputs, establish financial projections for cost in an IPD agreement in a series of reports that flow up to an aggregate metric. All of this information is part of tracking progress towards the target cost. *(See Target Value Design for more information.)* At a minimum, the IPD team needs to track:

> Risk register for the project (*see Risk Management for more information) with current weighted risk;*

> Productivity (cost to complete) projection for each IPD team member; and

> Current contract values for all companies along with allowances, contingencies, etc.

The IPD team should take the total of all current contracts and allowances, then add the net of the current weighted risk on the risk register and the sum of each IPD team member's firms' productivity projections to arrive at the current projected final cost on the project.

This number is compared against the contract terms to determine whether the IPD team is using contingency, is over budget, or projecting a savings.

It is important to set up a system in which nobody wins if other parties lose. This means that one's success depends upon the success of all.

How do you ensure financial accuracy and save on costs?

Give people on your team time and education to get up to speed. Financial transparency can be difficult for some members of the IPD team, as they are not used to opening their books to others and allowing others to see what they are being paid and what others on their team are being paid.

Establish a process to develop cost estimates. Create and establish a process with your IPD team to develop cost estimates at the different stages of design. Discuss what these processes will look like, and come to a shared understanding about how they will be used on the project and how they will help to inform project outcomes. Different amounts of detail and speed are needed depending on how the cost estimate is going to be used. A change-order estimate may require a significant amount of detail, while a single rough order of magnitude may be all that is needed for the IPD team to make a decision between competing design solutions. It is okay to wait weeks for detailed estimates in some cases, while in other cases, the IPD team may need an idea of the cost estimate by the end of the day.

The IPD team needs to understand the difference in confidence level with an estimate that is produced in a day with little detail versus a detailed estimate produced with detailed design information and subcontractor input over a period of a couple weeks.

Use conceptual estimating. In the early phases of the design of a project, conceptual estimating is an extremely important skill set to have on the project team. Project teams should be able to discuss high-cost impacts of design decisions without a significant amount of detail in the design. This means that an estimator will be asked to provide cost estimates without a detailed set of plans to use for a take-off estimate. This can be very challenging for estimators who are not used to this kind of a process.

Track productivity and burn rate. Track productivity and burn rate from the *start of design* and through construction every week. Check the burn rate against your cost-loaded schedule and target schedule by collecting weekly trade partner time tickets and material receipts. Be disciplined in your tracking. Schedule financial report outs, with all project team members reporting on actual expenditures, productivity, and alignment to forecasts. Tracking against a

DEVELOP A TEMPLATE FOR COST TRACKING

Often, teams on IPD projects end up reinventing the wheel when it comes to measuring and visualizing cost data. One way to prevent this extra work is to develop a template that you could continuously use across projects for cost tracking.

Financial transparency can be difficult for some members of the IPD team, as they are not used to opening their books to others and allowing others to see what they are being paid and what others on their team are being paid.

budget projection will allow the IPD team to adjust to control spending if possible. Confront slippage in the budget immediately. If the design is executed inefficiently, the IPD team's profits could be depleted before construction even starts. The most common way a firm will go over budget on a project is by spending more each month than budgeted in their estimate. This is particularly common for design firms. *(See Appendices 23 and 24 for spreadsheets used to track projected and actual costs over time. See Appendix 25 to see an example of contingency tracking.)*

Reduce and show savings through shared costs. Some IPD team members will have redundant work programs or processes. For example, in a traditional project, three trade contractors working on a project will each plan for their own refueling program or cleaning crew. In an IPD project, these costs should be transparent and discussed. Frequently, IPD team members can share work and expenses to eliminate paying for redundant activities.

Have monthly budget reviews on actuals and forecasted costs. Bring the IPD team together during design and construction to monitor the budget, find trends in costs, and forecast costs on the project. Conduct monthly forecasting from design to construction with your team. *(See Appendix 22 for an example of a project financial update.)*

Train team members in cost tracking and forecasting. There can be a steep learning curve for many trade partners and other IPD team members to use metrics and forecasting at a high level of detail effectively and consistently. Invest time in training trade partners on forecasting and tracking. This will help keep everyone on track and will be an investment in your local industry.

Hire a third-party auditor to monitor costs. If monitoring and tracking costs is too much of a burden for IPD team members, consider hiring a third-party auditor to monitor costs on the project.

Related chapters include: **Validation** (pg. 63), **Target Value Design** (pg. 67), **Risk Management** (pg. 81)

LEAN THINKING
How Do We Maximize Value, Eliminate Waste, and Make Reliable Commitments?

What is lean thinking?

Lean thinking is about defining customer value, mapping the chain of value, establishing pull, creating flow, and finding the right problems to solve. Additionally, it focuses on developing people and consistent improvement, with the goal of waste reduction and value creation. There are many resources available through organizations such as the Lean Construction Institutes of the U.S. and Canada.

Why is lean thinking important in IPD?

In general, projects are composed of diverse groups of companies, people, and skill sets. Teams work well when they can become self-aware of their processes and reflect on their performance with a mind toward improvement. A focus on clearly defining value is important to set the team out in the right direction, which will allow the owner to receive the value. Streamlined processes allow for an efficient flow of information between team members, enabling the final delivery of a project with as little wasted time, effort, and money as possible. A constant drive to analyze and improve those processes allows a team to adapt and succeed in an ever-changing work environment.

What makes lean specifically important in IPD projects is that it creates a set of aligned financial terms between the owner and the design and construction IPD team members on a project. IPD's financial alignment exposes many traditional communication and planning practices as wasteful and inefficient. To maximize the value of the contract structure, teams require a new work philosophy focused on efficiency and reliability. Teams look to lean as a management system and set of processes and mind-sets to create a more efficient work system.

> IPD's financial alignment exposes many traditional communication and planning practices as wasteful and inefficient.

How do I apply lean thinking on IPD projects?

Define and document customer value.

To successfully deliver a project with minimal waste, the project team must clearly define the customer's expectations. To identify and document the owner's and other project team member's values for a successful project, you can use the following lean practices:

> **Validation Study**
>
> A Validation Study (or Validation Report) is a collaborative report that captures the owner's value proposition as defined by the final business case, budget, schedule, and program of a project. (*See Validation for more information.*)

The IPD team is included at the start of a project, often prior to finalization of the owner's business case, and works to develop the Validation Study. Understanding why a project exists prior to developing conceptual designs gives the team freedom to explore diverse options to deliver value. The completed Validation Study acts as a guide for the project team through the design and construction process to orient them back to the owner's value proposition. *(See Appendix 8 for an example of a validation checklist used to build a Validation Study's table of contents.)*

Set-Based Design

Set-Based Design is a concept of advancing multiple designs to make the best decision established on additional information gained from further design development. This results in the team having significantly more information during final decision-making. For example, a team may advance three structural systems into the Design Development phase along with floor plans and shaft layout.

A3 Thinking

An A3 is a structured process of documenting a problem, options, proposed solution, and an action plan on a single sheet of paper (A3 refers to a standard 11" x 17" sheet of paper). This process first finds consensus around a problem statement and, in turn, helps to build consensus around a proposed path forward. A3s are developed collaboratively with all project stakeholders. *(See Appendix 12 for an example of an A3.)*

Choosing by Advantages

Choose by Advantages (CBA) is a systematic decision-making process that focuses on the advantages of options. This process is particularly useful when trying to reach a consensus in a large group of people with different goals and values. Unlike systems focused on assessment based on pros and cons, CBA recognizes that cons could also be an advantage for one or more of the other options.

Use streamlined processes.

Unlike traditional processes for communication and accountability that contain inherent waste, lean processes help project teams level workloads and reduce waste. Lean streamlined processes include:

Last Planner System

IPD teams use the five connected conversations of the Last Planner System (LPS) to manage activities from early feasibility studies through construction and commissioning. Projects are started with high-level milestones, then phase pull plans are created as the work proceeds. Look-ahead planning, weekly work planning, and learning (measured through percent plan complete and variances) are implemented to manage the weekly work of the team. *(See Appendix 16 for examples of weekly work plans.)*

Pull Planning the Design and Construction Work

Pull planning is a component of the LPS and a process many teams use to start their lean implementations. Pull planning is a powerful way to get designers and trade contractors to make commitments and help schedule the design and construction work. Project team members start with a milestone and then work backward logically to determine all of the steps needed to complete the work and all of the dependencies between the steps. These processes can break down communication barriers and build collaboration between team members.

Co-Location

Co-location helps align teams and increases collaboration by having all stakeholders work in the same office. While large IPD teams can co-locate for extended periods on a project, smaller projects can co-locate for shorter sessions (e.g., one day every two weeks) or by using online tools. *(See Co-location in a Big Room for more information.)*

---

5 See the website of the Lean Construction Institute for more information: https://www.leanconstruction.org/.

❯ Building Information Modeling (BIM)

BIM is used during design and construction for coordination, prefabrication, scheduling, cost estimating, and facilities management. *(See Integrating Project Information Using Building Information for more information.)*

❯ Information Management

To prevent waste in the flow and control of information, have the team develop, document, and display their processes for using and sharing project information. Implement a central storage point for each type of project information (e.g., using a cloud-based documentation platform and a systematic naming structure).

Encourage and commit to continuous improvement.

Successful, high-performing teams remain committed to continuous improvement on a project. Teams need to be self-aware of their processes and be able to reflect and improve processes when there is breakdown. Use the following lean tools to encourage and commit to continuous improvement:

❯ Plan-Do-Check-Adjust

Plan-Do-Check-Adjust, also called Plan-Do-Check-Act (both referenced as PDCA) is a four-step process. It creates a feedback loop for teams to assess their ability to achieve specific outcomes. This process is implemented with an expected outcome, which is then measured against the team's actual outcome. If there is variance between the expected and actual outcomes, the team conducts a deep dive to uncover any issues. During the deep dive, the team will also develop countermeasures to integrate into the revised process.

❯ 5 Whys

The goal of the 5 Whys process is to discover the root cause of why a team did not achieve an expected outcome. This process involves asking "why?" five times, each time drilling into why the previous activity occurred. Once the team discovers the root cause, it should find countermeasures to remove the root cause, helping to resolve similar issues in the future.

❯ Plus/Delta Thinking

Plus/Delta is a method of assessing what has or has not gone well on a project. A Plus is something that went well and should be repeated, while a Delta is something that didn't go well and should be improved. To conduct a Plus/Delta, have the team take five minutes, or less, to list Plus/Deltas at the end of each meeting or event. Next, have the team assign each Delta to a specific individual to resolve, with an action plan and commitment for completion.

❯ Conduct mid-construction and end of construction retrospectives

To assess progress during construction, conduct a mid-construction retrospective to see what went well, what did not, and to make any identified countermeasures to the construction process. Use surveys as a tool to conduct these retrospectives and receive team feedback about how they feel the project is going and to help develop countermeasures to project problems. At the end of the warrantee period, have your team look back at the entire project for an end-of-construction retrospective and see what went well and where there could have been improvement.

For more about these concepts, please review the More Resources section of the guide.

Related chapters include: **Validation** (pg. 63), **Co-location in a Big Room** (pg. 71), **Integrating Project Information Using Building Information** (pg. 79)

**PATH TO
CONTRACT**

Owner Alignment

Team Selection

The Contract Workshop/Team Alignment

**ONGOING
CONSIDERATIONS**

Team Management Building and Managing a Successful IPD Team

Financial Organization and Financial Monitoring

Lean Thinking

EARLY WORK

Validation: Go/No Go

Target Value Design

Co-location in a Big Room

Design Management

Prefabrication

Integrating Project Information Using Building Information Modeling

Risk Management

Project Dashboards

LATER WORK

Team Maintenance

**WHAT GOES
WRONG**

Closeout

EARLY WORK
Processes and Tools

———

IPD, perhaps more than any other delivery model, is heavily loaded on the front end in terms of effort. Teams find that enhanced project outcomes are best supported by laying rich foundations of project data and information, management and decision-making processes, communications protocols, financial controls, risk-mitigation strategies, and performance metrics.

After project initiation, the team moves through a robust validation process to test possible project outcomes against the owner's business-case objectives. To be successful, this early validation work on a project requires cultural and group process efforts, as well as the application of a range of specific tools and techniques.

Teams also grapple with defining how they will work together, establishing organizational structures as well as establishing expectations for how they will work together in a collaborative Big Room environment. Understanding the role of design and managing that process is particularly important in early team efforts.

Target Value Design is the process employed by the team to move from inception of validation to completion of project documentation, allowing the team to optimize value through collective creativity while developing both budget and solution with a methodical approach. A risk register is typically deployed to help identify, quantify, and mitigate risks proactively across the project. Project dashboards are deployed to monitor alignment to project-success criteria.

Published research outcomes and experiences shared by multi-IPD project participants underscore the importance of this early work. Understanding and embracing the front-end-loaded nature of IPD will set the project up for success.

VALIDATION
Is the Project a Go or No Go?

What is validation?

Validation is a process that establishes certainty for the IPD team and for the owner: it proves or disproves whether the IPD team can meet the full range of the owner's Conditions of Satisfaction (CoS) within the owner's allowable-cost and schedule constraints. Validation is *not* compressed schematic design: the project design is developed only to the degree necessary to achieve certainty. Validation is a go/no-go gate, undertaken at the beginning of the project and often has its own budget, schedule, and approvals. A number of prerequisites are required before validation can begin.

For owners, the main audience for validation will likely be internal, such as a committee or board that approves funding. However, validation should not be limited to an internal owner process since it is critical for establishing the IPD team's confidence about meeting the goals within the project's parameters. The outcome of validation (a validation report or other documentation) should be a touchstone for the duration of the project.

Why do validation?

The purpose of validation is certainty. The validation process results in a comprehensive report that is essentially a collective statement by the team: "We can build this building, that does these things, for this much money, in this much time." If the specifics of those outcomes are acceptable to the owner, it allows the owner and the team to proceed with confidence that the project is viable. Alternatively, if the outcome of validation demonstrates that CoS cannot be achieved within allowable cost and schedule constraints, the owner can make an informed decision about project direction (e.g., no go, change in scope, increase budget) for a significantly smaller investment than might be required in traditional models to achieve a similar degree of certainty.

The purpose of validation is certainty.

What do I need before I begin validation?

Find out the facts. Determine what information you have and can collect. Site information, feasibility studies, market need, labor market, regulatory and compliance requirements, other legal concerns, owner's business-case objectives, project history to date, preliminary program—basically, any existing information or facts that might influence the outcomes of the project—should be gathered by and for the use of the team during validation.

Ask, "What are the parameters of validation?" You need to know what exactly you are validating (e.g., scope, goals) and what you are not (e.g., long-term return on investment). Determine who should be involved (this can help set the validation budget), the approval cycle for the owner's audience (this can set the validation schedule), and whether there is a standard process used by the owner.

Determine the outcome of validation. Ask, "What does 'done' look like?" to help the IPD team optimize their efforts during the validation phase. Decide on the format, level of expected detail, required information, a checklist of things that will be addressed, table of contents or deliverables of a validation report, and other content. *(See sample checklist used in a validation report in Appendix 8.)* Always look for the minimum investment of effort and resources on the part of the team to yield the confidence necessary to make the collective commitment. This preliminary pass at validation outcomes will likely evolve over the validation phase and should be continuously reviewed.

Plan the logistics of validation. Planning the logistics of validation will help establish the budget and schedule for validation. Ask yourself, will there be a Big Room, and do you need a facilitator? Do you need a commonly used software platform for this phase, and what communication protocols will be used for this phase? Decisions here shape the information flow for data throughout the job. You need to consider ease of use, version control notification, live editing, shareability, software compatibility, security settings, achievability, change tracking, and the like. Using the desired outcomes of validation as a framework, determine milestones, tasks, and durations. Then pull plan tasks back accordingly to set the validation timeline. *(See Design Management for more information on pull planning the design process.)* Having a clear plan and estimate for validation enhances chances for a successful outcome.

What happens during validation?

The team takes all available information and begins exploring solutions, following the validation timeline. This is a combination of work both in and out of the Big Room, both in small interdisciplinary teams, such as PITs, and as individual stakeholder contributors in a mix of charrette-like activity and more traditional work. *(See Team Management for more information on PITs.)* Designs begin to evolve and information organized, all structured and tackled with an eye toward the end goals of the validation process (i.e., what "done" looks like). Remember that validation is not compressed schematic design—the end goal is certainty. The team should constantly be looking for the most effective way to achieve the necessary certainty with the minimum amount of investment of time and resources at this phase.

Things to consider during validation

Understand the owner's business-case objectives. The clearer the IPD team's understanding of the "why" of the project for the owner (and their criteria and processes for funding), the greater their ability to make recommendations for maximization of value. The more robust the set of project goals or CoS, the more likely the team will be able to shape a project that balances them or make clear the necessary trade-offs required to optimize some goals or CoS over others. This clarity greatly assists in managing and aligning owner expectations and outcomes. Understanding the owner's funding processes will also be important. (The

WHAT IS VALIDATION?

The Validation phase of an IPD project yields collective certainty and commitment to function, cost, and schedule on the part of the team: "We can build this building, that does these things, for this much money, in this much time." The goal of the team during validation is to develop the minimum amount of information necessary to make that statement with confidence.

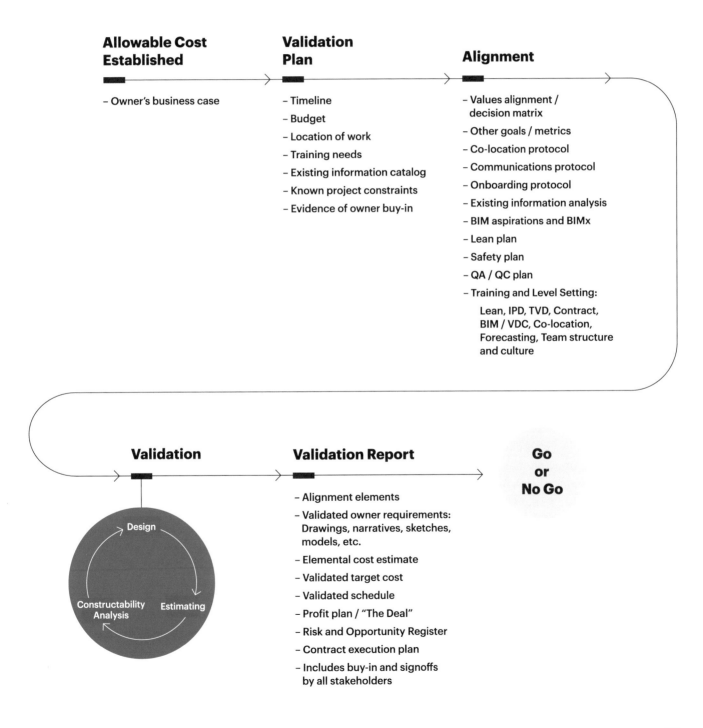

Allowable Cost Established

– Owner's business case

Validation Plan

– Timeline
– Budget
– Location of work
– Training needs
– Existing information catalog
– Known project constraints
– Evidence of owner buy-in

Alignment

– Values alignment / decision matrix
– Other goals / metrics
– Co-location protocol
– Communications protocol
– Onboarding protocol
– Existing information analysis
– BIM aspirations and BIMx
– Lean plan
– Safety plan
– QA / QC plan
– Training and Level Setting:

   Lean, IPD, TVD, Contract, BIM / VDC, Co-location, Forecasting, Team structure and culture

Validation

Design
Estimating
Constructability Analysis

Validation Report

– Alignment elements
– Validated owner requirements: Drawings, narratives, sketches, models, etc.
– Elemental cost estimate
– Validated target cost
– Validated schedule
– Profit plan / "The Deal"
– Risk and Opportunity Register
– Contract execution plan
– Includes buy-in and signoffs by all stakeholders

Go or No Go

High value design can save on constructibility costs and provide efficiencies in operations, saving money on staffing.

following examples in the Appendices illustrate how CoS might drive decisions. Appendix 19 shows how a team determined what to use as key performance indicators. Appendix 9 is a decision tool evaluating a decision's impact on the CoS values.)

Define the budget initially as an owner constraint, not a target. Owners will have limits to how much they can spend on a project (i.e., the allowable cost based on a business case). At first, define this budget as an owner constraint or boundary condition, rather than a target for the project. As validation proceeds and approaches conclusion, the relationship of the budget to a target will become clear, with the team generally setting target cost at the end of validation.

Avoid rigidity with costs. Sometimes IPD teams and owners can be rigid about costs, looking for precise accuracy in the numbers. This can trigger the need to generate more detail than is helpful, drawing out the process longer than necessary. Have estimators on the team who are skilled and comfortable in the realm of conceptual estimating, making and documenting assumptions in concert with designers in an iterative design-like fashion. As the design evolves, the detail in the estimates will evolve, and vice versa.

Define the scope of the project. Often on projects, people skip over scope in the validation process. However, you cannot validate costs with confidence without defining the scope of the project. Have clarity and agreement on the scope and program with the team, whether developed during validation or before. Document assumptions.

Understand the value of design. The validation process is largely an exercise in design and is one of the IPD team's most significant opportunities to create value for the owner's enterprise. High value design can save on constructability costs and provide efficiencies in operations, saving money on staffing. Investments here will yield significant paybacks; validation should not be skimped on or rushed.

Ensure comprehensiveness and clarity of the validation report. The validation report should be thorough enough and clear enough that a reader not familiar with the project could arrive at a similar degree of certainty about the outcomes as the team.

Related chapters include: **Design Management** (pg. 73), **Team Management** (pg. 41)

TARGET VALUE DESIGN
How Do We Set and Achieve the Right Target Costs?

What is Target Value Design?

Target Value Design (TVD) is a philosophy of designing to a budget instead of budgeting a design. Cost estimating becomes a crucial part of the development of a design with constant checks against the target budget. Therefore, the goal is to design to a detailed budget, as opposed to waiting and budgeting a detailed design. Target cost is established during or after validation and TVD is the process that helps the team design to that target. *(See More Resources for detailed information.)*

Why is Target Value Design Important?

Traditional projects estimate costs at design milestones, such as 100% Design Development or 50% Construction Documents, and estimates can greatly increase at each check-in point due to the large number of design decisions documented on drawings between milestones. TVD provides near-immediate feedback to the design team on a continuous basis to better inform decision-making. This feedback allows the team to maintain the project budget as the design develops from early conceptual plans into buildable documents.

Who needs to be involved in Target Value Design?

You need to have the right people at the right time to drive TVD. Achieving TVD requires extensive collaboration between designers and builders, particularly cost estimators and trade partners. The firms who will build the most complicated pieces of the project should be at the table early to provide continuous feedback as the design progresses. This allows trade partners to influence the design in a way that achieves the owner's goals while maintaining the budget.

The goal is to design to a detailed budget, as opposed to waiting and budgeting to a detailed design.

How do you do Target Value Design?

Set targets. The target cost and value conditions are established during or after validation. Typically, this is a dollar amount determined by the project management team and senior management team groups based on some prior experience or market data, but for some owners, cost may be less important than return on investment or program needs. Whatever the target, it can be based on previous projects, cost per square foot, a third-party estimator, or other means. Most often, the target is an aggressive "stretch" goal. If this is a cost goal, it would be set below market but within reach, if the team can work together.

Divide the budget. Divide the budget into smaller, anageable pieces and assign a project implementation team (PIT). Next, the budget will be divided into manageable sized pieces and assigned a PIT. The divisions will vary by team, but typically PITs organize around construction trades or elements that require a great deal of coordination to design and construct. *(See Team Management for more information on PITs.)*

Set subtargets. Set a subtarget for each PIT, knowing that all the PIT subtargets will be considered together in order to achieve the overall target for the project. Each PIT reports out on opportunities and risks to their piece of the budget as they track progress against their subtarget. This process allows for a give and take between the individual members within a budget PIT. The aggregate target for all the elements in the PIT is what is important, not the final value of any individual contract for PIT members. *(See Appendix 26 for an example of how an IPD team tracked PITs in TVD.)* The individual decisions within TVD should be tracked on a risk register. *(See Risk Management for more information.)*

Assign PIT leaders. Assign a leader for each PIT to track the individual items under discussion and to report out to the project leadership team about the progress of their PIT against their individual target. Since a risk or opportunity identified by one group may impact another group, the PIT leaders should regularly confer. For example, to save costs installing a precast exterior panel, the enclosure PIT may propose a structural kicker. However, while this change may be cheaper for the skin, it may cost more for the MEP design and impact MEP coordination. On an IPD project, the skin PIT would need to make sure to ask the MEP PIT, "How does this change affect you?" Then, have discussions about overall impact on the project prior to finalizing their design.

Other things to consider in Target Value Design

Focus on the process to understand cost drivers. The process of understanding cost drivers is one of the early steps to driving down the cost of the project. Focus on the smallest percentage of items that create the largest percentage of the cost. Once a project team understands what is driving the bulk of the cost on the project, they can find ways to minimize the risk and lower the cost to deliver the project.

Set aggressive targets to help team innovation. Setting aggressive targets will stretch the project team to think about how projects are delivered. This will push the project team to think creatively to find innovative solutions that will help them achieve those targets.

INNOVATIVE IDEAS IN TARGET VALUE DESIGN

One innovative way of estimating for TVD is to address the function and use of the building. This is the method that the Finnish firm Haapola uses for estimating. The firm engages the owner prior to validation. The firm then collects function and utilization information from an owner and uses that information to help the owner make better, higher-value design decisions. Using this information, they establish the target or the allowable cost based on their function and use analysis, eventually leveraging this data to build cost models. This data is then provided to the team, and the firm works together with the team to arrive at the best design and construction solutions.

Be flexible with subtarget goals. Different trades and PITs may not be able to hit the same percentage saving targets. For example, as a percentage of the PITs value, Site and Core/Shell PITs typically recognize a higher percentage savings than MEP PITs. Ask different PITs and trades to set and manage individual targets that equal the overall budget target.

Use lean in design to increase information flow. Lean in design uses processes aligned with design thinking to explore multiple options in parallel with clear criteria. It provides more information to designers, which, even if it takes more time, results in better value for the owner. Increasing the information flow between contractors, estimators, trade contractors, and designers means that designers have better information on pricing, products, and constructability.

Use skilled conceptual estimators or technical experts to help with design decisions. A skilled conceptual estimator or technical expert knows what kind of design decisions will drive a cost up or down. This is a different skillset than traditional estimating, and care should be taken to make sure the project team has a right fit for this role. A team may also include skilled technical designers who understand constructability and cost impacts of design decisions. Ideally, both of these should be brought to the team to work together for best results. Estimators and designers should work closely throughout.

Encourage estimators to provide a cost range. Estimators often use a very precise process to determine and make assumptions about cost. For TVD, have estimators work a cost range rather than a precise number. Project teams can use this information to explore innovative approaches during the TVD process.

You need to have the right people at the right time to drive TVD. Achieving TVD requires extensive collaboration between designers and builders, particularly cost estimators and trade partners. The firms who will build the most complicated pieces of the project should be at the table early to provide continuous feedback as the design progresses.

Related chapters include: **Team Management** (pg. 41), **Risk Management** (pg. 81)

Considerations	Characteristics
Attributes	– Bright – Flexible – Comfortable – Accessible – Safe
Environment	– Natural light – Lighting – Ventilation – Temperature – Acoustics
Location	– Proximity to team offices – Proximity to site
Safe / Safety	– Access – Parking – Lighting – Inclement weather protection
Comfort / Convenience	– Washrooms –Kitchen Access – Coats / Boots
Functionality	– Shape of room – A/V – Internet – Phone – Projection – Wall space – Furniture / furnishing / layout / adjacencies – Storage – Marker walls – Parking – Access

CO-LOCATION IN A BIG ROOM

*Why are We All Together and
How Do We Make the Most of It?*

What is co-location?

Co-location is when team members from different firms meet and perform work together in one physical space over an extended period of time on a project. This strategy can significantly improve collaboration and help build team relationships. Remember, co-location is working side by side, which may happen in a Big Room or other (even virtual) spaces. Simply holding meetings and having materials posted in a Big Room is not the same as co-location.

Why is co-location important?

For any design and construction project, the sheer magnitude of participants, information, analysis, and handoffs, coupled with the pace at which decisions and actions are needed, present formidable challenges. Co-location helps the project team overcome these challenges, allowing efficient sharing of pertinent project information; reducing latency in asking and answering questions; supporting collaborative work on innovative project solutions; eliminating misunderstandings, rework, or gaps in expectations; and expediting project decision-making.

Should I co-locate my team?

Deciding whether to co-locate your project team depends on the project team and project's needs. Co-location can be dictated by the IPD experience of a team. For participants new to IPD, co-location can help change traditional ways of working and communicating together. Working together in the same room is an effective way to break down traditional disciplinary silos.

Co-location also depends on the project: project complexity and scope may inform the need to co-locate specific team members. Small projects can be challenging to co-locate, as they usually do not require team members to work on the project full time. A project with a budget of $10M may be better served by virtual co-location with limited physical co-location. Conversely, a project with a $250M budget might benefit from full-time face-to-face co-location during design and construction. Many projects use a one-, two-, or three-day-per-week co-location strategy, with project team members working from their offices on other days.

Simply holding meetings and having materials posted in a Big Room is not the same as co-location.

Who should be co-located?

The right people need to be in the room for the decisions being made and the work being performed. This may vary over time and is informed by the work plan and agendas set by the IPD team. It is important that co-located team members are a good behavioral fit. Ensure that there are also participants in the room that have the authority to make and execute decisions (e.g., project management members).

How do I co-locate my team?

Provide the team with appropriate amenities, space, and resources for co-location. Co-location requires space for everyone to work, amenities (e.g., a place for lunch, break-out spaces, large group meetings), and other resources (e.g., computers, software) for proper room setup. Determine whether the co-location space will be a cost or possibly be provided by a team member with space they might "loan" to the team.

Select the right scale of engagement for project team members. Project team members, especially those who are not on the IPD team, can become frustrated if they are asked to co-locate for more time than needed to do their work or for insufficient time to help the project team make decisions. Be realistic about how much time project team members will be needed on the project from week to week and scale up or down as needed. Review the co-location plans on a regular basis. Be mindful of the impact of co-location on project team members' companies—companies of smaller size or with smaller scope may be challenged more than large ones. Presence in the co-location space to share expertise and experience on demand is the most important aspect of co-location. Therefore, it is fine if project team members are in the space working on other projects.

Determine how long to co-locate on the project. Schedule how long you expect the project team to co-locate on the project and who is required to co-locate. Have project team members buy-in for the duration of the co-location schedule and stick with those periods of duration—consistency of presence is critical. If the co-location period extends longer than expected, it could harm trust between team members.

Establish rules of engagement. Big Room rules of engagement help provide structures around the duration and frequency of project team interactions. These rules allow you to respect the working times of project team members while allowing for enough time for the project team to react to project challenges.

Encourage all project team members to work together and listen to one another. Designers and contractors need to have strong communication skills and be willing to work as an integrative team.

Celebrate success! To encourage a positive team culture, owners should budget for celebrating the successes of the project team as they achieve specific milestones. These celebrations can be informal in nature, such as having a barbecue for the project team.

Create a learning organization. Make sure your co-located project team members continue to acquire new skills and knowledge. For example, you can educate the project team in lean concepts, or form your own community of practice to learn from one another as the project moves forward. Take time to learn and develop—treating the project team like a functional organization intended for the long haul and making time for learning will pay dividends.

RULES OF ENGAGEMENT

On one project, the architects co-located several days a week with the contractor. However, as the project was nearing construction, the architects were receiving so many questions throughout the day about the design that they fell behind their design schedule. The project team then set up rules that provided set Q&A times in the morning and afternoon when the architects would be accessible for project team design questions.

DESIGN MANAGEMENT
How Do We Collectively Leverage the Power of Design?

___

What is design management?

Buildings do not come into being without a reason; rather, an owner has a value-generating enterprise they need to support. The design phases are the project team's greatest opportunity to shape and impact the supported activities and to maximize value for the owner's enterprise. Design management is about supporting and managing the design process toward optimized outcomes, both from a design process and project perspective.

Why does design management matter?

While there is a long history and science to construction management, design management traditionally has lacked a similar structure and rigor. For many architects, the design never ends, even well into construction. The act of design is iterative in nature and distinct from the linear nature of construction. Designers do not know what they do not know until they find it, creating an inherent level of uncertainty to the work. However, IPD requires that design is also predictably planned. This is because there are more people involved in the collaborative process: completing one's work depends on everyone else completing and sharing their own work in a timely fashion. Appropriate design-management techniques help the design team focus on exploration and documentation that yield the highest possible returns in a timely, reliable fashion.

How is the design process managed?

At minimum, project teams should consider at least three things when looking at managing design: defining what "done" looks like before starting design, planning, and decision-making and information flow. These topics are deeply interconnected.

The design phases are the project team's greatest opportunity to shape and impact the supported activities and to maximize value for the owner's enterprise.

What Does "Done" Look Like?

If project teams think very critically about what "done" might look like before they begin design activities, they can avoid redundancy, maximize value, and minimize waste. "Pulling" requirements from builders to help decide what "done" looks like before starting design tasks will help identify missing and necessary information inputs, ultimately allowing designers to better

Focus on and describe the quality and handoffs and deliverables.

focus on exploring a more targeted range of alternatives and maximizing return on design investments.

One goal for design on IPD projects is to only create deliverables that provide the necessary information to support decisions and construction in the most efficient possible form. Effective and efficient forms of deliverables may include a spreadsheet, a narrative plus a vignette, a 3-D detail, a vendor cut sheet, or a shop drawing. With traditional contractual boundaries between stakeholders removed on IPD projects, information packaging can become richer, significantly streamlined, and better fit for the purpose.

One tool that can be quite helpful in defining what "done" looks like is a BIM Level of Development (LOD) matrix. *(See Appendix 13 for an example.)* Completing a LOD matrix (and going through the necessary conversations to do so) will help the project team clarify and define project scope, and collaborate more effectively. This matrix will also aid in defining roles and responsibilities for different areas of design, setting design milestones, tracking work progress, and informing pull planning for design. Integrate the LOD matrix into your delivery schedule. (For more information on using a LOD matrix, see Integrating Project Information Using Building Information Modeling.)

Sometimes there is a tendency to overdevelop the design model, leading to inefficiencies in work hours. Have the project team talk collectively about what level of detail is most appropriate for specific project deliverables, as well as which stakeholder might be best suited to deliver each level of development, always keeping an eye toward the minimum effort necessary to capture necessary information. For example, if a trade contractor must detail to produce fabrication models or documentation, you can prevent duplicated efforts by ensuring that the design team does not detail the same work that the trade contractor will require later in the project. Make sure the project team has the right people in the room when completing the LOD matrix, including the specifications writers.

How Do I Plan during Design?

Pull plan the design process. Pull planning during design is different than pull planning during construction. Design, unlike construction, is a nonlinear, exploratory process. During the pull plan meeting, first define what "done" looks like: what are the end decisions or deliverables? Next, identify design milestones (e.g., necessary decision points or major information handoffs; specific deliverables, like programming or envelope; or major project events, such as necessary approvals). Pull tasks back from "done" to and through milestones, focusing first on sequence. Be cognizant that iterative loops in the plan may occur. This is part of the nature of design and should be expected and accommodated by the project team. Work to frame tasks in terms of necessary decisions or information more than deliverables. Once the sequence has been satisfactorily defined, attach durations and hours estimates, then review. It is likely that the process will take several iterations to achieve alignment between milestones, schedule, and budget. It is important to have the right people in the room to determine both sequence and duration—those who will be doing the work should plan the work.

Hold regular work-planning meetings. Determine the frequency of meetings needed to plan and to check on progress and direction for the design process (generally weekly or biweekly, at most). Have designers create clearly defined weekly work plans with associated deliverables (e.g., information, decisions, or documentation). Track tasks and completion using percent plan complete. These regular measurement and planning meetings will provide early warnings if things are not going according to plan, allowing the project team to better adjust to stay within target budgets—they also support a culture of reliable commitment. If the project scope and co-location situation warrants, daily huddles with project team members can be helpful to ensure that the project team is on track with their work commitments. Daily huddles also assist in building a strong and trusting team culture.

Plan decisions and deliverables early and continuously as a team. Review the overall pull plan at every milestone. This will help the project team track design work across the project, keep people focused, and prevent the project team from going over budget. Use the risk register as a checkpoint to make sure risks are being mitigated (*see Risk Management for more information*) and to find potential opportunities for betterments.

How Do I Plan for Decision-Making and Information Flow?

Map out information flow during design, identifying required formats and schedules. Project team members need to be able to share information that is meaningful and trust that information will arrive on time. Designers have different needs from contractors and other project team members in how their information is formatted and when information is needed for decision-making. Focus on and describe the quality of handoffs and deliverables. Ensure that everyone on the project team understands the information needs and expectations of others to help map out information flow during the design process.

Focus on commitments and decision-making. It can be difficult to receive good reporting on productivity measures during design. Focus on defining commitments and decision-making, and use these to help with forecasting.

Select lean processes that best fit design decision-making. A3s and Choosing by Advantages are powerful processes for improving decision-making. However, there is a learning curve, along with different costs and regional expectations around which tools to use. Some will fit better than others for specific types of decisions or for specific types of teams. Choose the lean process that best fits your needs in design.

Other considerations during design

Always ask whether the right people are in the room. Design in IPD requires extensive collaboration between different disciplines. Ensure that you have on-boarded the contractor and trade partners to enable them to be a part of the design process. Only select people needed for specific meetings or co-location. The input of some team members may easily be provided via email or by phone.

Decide on co-location for design phase. Trade contractors working next to designers is a significant benefit of IPD. The collaboration between designers and those actually building the components is key to getting the right design. Determine whether your project would benefit from co-location. (*See Co-location in a Big Room for more information.*)

Conduct mid-design retrospectives. Lessons learned are often developed at the end of a project. While useful for the next project, this does not help to correct issues during the current project's design. To assess progress during design, conduct a mid-design retrospective to see what went well, what did not, and to identify countermeasures or corrections to the design process.

Track betterments throughout design. Betterments are elements that do not fit in the current budget but could be implemented in design if there is enough project savings available. Have the team track opportunities taken and actual cost reductions and innovations

ELIMINATE THE FIGHT OR FLIGHT INSTINCT AND MAKE IT POSSIBLE TO BE CHALLENGED

When challenged on an idea, there is a desire to defend one's idea at any cost, which can halt productive discussion. One contractor suggested that you can intentionally shift out of having a fight or flight reaction by coming in with the right attitude. You can say to yourself, wait a minute, as stupid as that sounds, they just might be right.

One goal for design on IPD projects is to only create deliverables that provide the necessary information to support decisions and construction in the most efficient possible form.

made. Concurrently, track project savings to determine whether implementing a betterment will be cost effective for the project.

Establish project dashboards. Project dashboards should visually track cost, the schedule, team behavior/performance, quality improvements, and productivity improvements. It can help the team discuss the status of the project, as well as identify potential problems. (For more information, see Project Dashboards.) *(See Appendix 14 for examples of dashboards used as visual management in a Big Room.)*

How Do I Manage Design Budgets?

Have team members provide a staffing plan. To help determine and track costs during design, have team members create a detailed staffing plan that identifies the resources and staff they need to complete the project. Have team members track their staff numbers on the project to identify necessary changes to the plan. Make these changes explicit, and discuss them during weekly planning meetings and monthly forecasting.

Conduct a review at monthly forecasting meetings. Identifying overages in design and ensuring financial accountability between project team members require that the design team provide their time and cash-flow information. During monthly forecasting meetings, the project management team can conduct a time and cash-flow analysis to identify overages. Weekly work planning sessions will provide early warning signs. Should schedule challenges or overages occur, the project team should set corrective action plans.

What other tools can be used to manage design?

Consider including the following in your tool kit for design management:

> Responsibility matrix *(see Appendix 6 and 7)*

> Budgeting metrics *(see Appendices 22, 23, 24, and 25)*

> Scheduling metrics *(see Appendices 14, 15 and 16)*

> Risk register *(see Appendix 10)*

> Assumption logs

> Definition of the decision-making process

> Assessments to monitor team culture and alignment to goals and values *(see Appendices 19, 20, and 21)*

> Design-structure matrix

Related chapters include: **Integrating Project Information Using Building Information Modeling** (pg. 79), **Risk Management** (pg. 81), **Co-location in a Big Room** (pg. 71), **Project Dashboards** (pg. 83)

PREFABRICATION
Will Off-Site Construction Pay Off?

___

What is prefabrication and deliberate building assembly planning?

Prefabrication, or off-site construction, is the purposeful planning, design, and assembly of building elements at a location other than where they will be finally installed.[6] Prefabrication is well suited to IPD since these projects require early alignment of the project team, providing the team the opportunity to plan the project scope and budget before design begins. Therefore, IPD enables project teams to consider how to construct the building to achieve the best value for the owner, including whether the specific components, assemblies, and modules could be prefabricated prior to their arrival at the construction site.

How does prefabrication benefit an IPD project?

Prefabrication is a strategy that mitigates uncertainty and variability on a project while providing the project with multiple enhancement opportunities. When integrated teams plan fabrication early and deliberately, they can improve the schedule through concurrent construction, improve quality and labor productivity by using shop labor (which has qualified crews doing consistent work), increase safety and reduce risk, bring reliability to project costs, and eliminate non–value added steps in the construction process.

___

6 Off-Site Construction Council, "About the Council," *National Institute of Building Sciences*, accessed 3/2/2018, http://www.nibs.org/?page=oscc.

How do you use prefabrication effectively?

Start early. To get the maximum benefit of using prefabrication on the project, discuss fabrication methodologies, opportunities, and risks early. These conversations should begin when considering and interviewing the contractor and trade partners. Seek out opportunities where design-to-fabrication processes will allow for specification and detailing to happen once the project is at the detailer and fabrication level. Utilize a Level of Development (LOD) matrix to define for the builders what they need to build and how it is being drawn, specified, and detailed to incorporate design and fabrication accordingly. (For more information on using a LOD matrix, see Integrating Project Information Using Building Information Modeling.)

To get the maximum benefit of using prefabrication on the project, discuss fabrication methodologies, opportunities, and risks early.

Ensure that you have the processes and logistics in place to commit to prefabrication. Prefabrication requires design decisions to occur early and at a higher level of detail than on traditional projects. Confirm that project design processes, design coordination, and decision-making capabilities are stable enough to commit to prefabrication. For example, the prefabrication of modular bathrooms requires early commitment to the following:

> Holding dimensions for structure;

> Assessment of slab depressions;

> Review of assessment determining whether usable space will be reduced by prefabricated walls meeting other walls;

> Agreement to limit the number of bathroom types and sizes; and

> Early decision-making on bathroom layout and finish selections.

Determine what to prefabricate. Consider prefabrication for repetitive elements of a building and complex elements that can be produced more accurately, productively, and safely under controlled factory conditions. These elements include:

> **Single-trade assemblies**
> Such as preassembled plumbing stacks for bathrooms or panelized exterior curtain wall.

> **Multi-trade assemblies**
> Such as above-corridor racks containing ductwork, plumbing piping, power conduit, and low-voltage cable trays. These assemblies can be delivered and erected in truck-length sections.

> **Modular components**
> Such as fully prefabricated bathrooms or classrooms fabricated from two fully finished modules that are joined on-site.

Establish the required quality of prefabricated elements. Be deliberate about the level of quality needed for prefabricated elements. Provide the off-site constructor with clear specifications as to the level of quality required. Determine if traditional specification requirements will govern prefabricated elements or whether the project team needs to define specific criteria to obtain the level of quality required.

Confirm responsibilities and liability at each stage. Determine which entity has responsibility and liability at each stage of fabrication, storage, transportation, installation, connection, and testing. You will also need to agree on the criteria and schedule for payment of prefabricated elements. Note that some contract terms may separate prefabrication costs from impacting the IPD team's profit pool. Note that significant modular elements purchased as products (i.e., not collections of materials and equipment fabricated with the labor of the project's trade contractors) may legally be considered as products. For example, in the United States, products are governed by the Uniform Commercial Code and product liability. In such cases, consult with legal counsel to ensure that appropriate responsibility has been placed.

Establish how fabrication and quality oversight will be verified, including quality assurance and quality control. Determine how the project team will confirm their adherence to quality assurance and quality control standards. This includes determining:

> How mock-ups will be reviewed and confirmed, prior to releasing for fabrication;

> Who will make inspection visits; and

> How will inspections be documented. Make sure to understand the submission requirements for regulatory officials, as well as their inspection requirements. Regulatory officials may be required by code to make off-site visits or may require third-party inspections and certifications. Build these into prefabrication planning and project schedule.

Related chapters include: **Integrating Project Information Using Building Information Modeling** (pg. 79)

INTEGRATING PROJECT INFORMATION USING BUILDING INFORMATION MODELING

How are We Leveraging the "I" in BIM?

What does it mean to integrate project information using BIM?

A building is a complex combination of systems developed by numerous parties over time. The integration of these systems into one 3-D source aids all members of the project team to visualize how systems interact, allowing them to identify gaps, clashes, and needs, and seek opportunities to innovate and improve. Development of the owner's program into documentation from which the project team can build the project is most effectively done through the use of a consolidated BIM, which is created, advanced, interrogated, and referenced by all parties.

Why is BIM important?

Using BIM for collaboration and 3-D interaction enhances the project team's ability to understand scope, to plan the schedule, and to verify cost, all of which increases certainty that the project will meet the owner's objectives.

How do you integrate project information with BIM effectively?

Have a robust discussion on BIM needs for the project. Early in the project, begin a robust conversation about the value of BIM and how to use it on the project. This conversation should occur no later than the contract workshop. During these early conversations, spend time determining BIM roles and responsibilities, including how to best manage collaboration when using the model. The discussions should also involve the management of technical concerns (e.g., finding the right points in space, where in the cloud the model will be stored) and of people (e.g., determining who will be the technical coordinator of the model), as well as deciding to what level design is developed at which stage of the project. (To learn more about a Level of Development [LOD] matrix approach, see Design Management.) Although this requires a high level of collaboration, identifying BIM leaders will maximize the effectiveness of BIM on the project. Ask yourself, will the project benefit from integrating the schedule into the model (4-D) and/or cost (5-D)?

Determine long-term BIM needs for the owner. Using BIM is a wonderful opportunity for owners and user groups to visualize, in advance, how they will use the space and to better understand their own operation and facility-management needs. It is also an opportunity to determine the convenience and safety of building maintenance for facility operators, from changing a lightbulb to maintaining major equipment. Owners may also require the project team to produce record models for facility operations and maintenance. To maximize the benefit for the owner, discuss with the owner their information needs for BIM. Ask whether the model will be used as is by the facility's staff. Determine whether the owner needs the entire BIM, or whether the owner prefers to select specific data from the model and other information sources to populate a facilities-management system and to provide visualizations (e.g., photographic or "light" BIMs) of complex areas, such as mechanical rooms. Based on the needs of the owner, plan in advance how to use the model and what information is required on the project. If the project incorporates renovation, laser scanning or reality capture (e.g., photos taken to cover a 360-degree view) of existing conditions can be merged with BIM to show new conditions simultaneously with existing. This is essential to visualize how to work with existing conditions, the sequence of work on the project, and how to get the renovation right the first time.

Determine what information you need to build.
Develop a LOD matrix concurrent with a design-phase pull plan and decision-making matrix for the project. *(See Appendix 13 for example LOD matrix.)* Consider the information you need to build. What information is needed for authorities having jurisdiction? What information does the owner require for user group and constituency approvals? Bring project parties together to develop a design-phase pull plan to establish the milestones, handoffs, and sequence of design development. Utilize the pull plan to identify milestones by which design elements need to reach a certain LOD. Draft an LOD matrix with design elements, milestones, the party responsible for developing that element, and the level of completeness required. This LOD matrix will become the road map to tracking decision-making and design progression. Develop a disciplined approach to reviewing and updating the pull plan, the LOD matrix, and design itself. Ask, are we achieving the milestones and at the expected LOD? If not, what are the consequences and options?

Develop a BIM execution plan to incorporate the LOD. Based on early conversations with the owner and project team about the value of BIM and their information needs, develop a BIM execution plan. This plan should define the roles and responsibilities of the project team members developing BIM.

Make information flow meaningful. Ensure the BIM execution plan maps out information flow during design, identifying required formats and schedules. Project team members need to be able to share information that is meaningful and trust that information will arrive on time. Project team members from different disciplines will have different needs for how their information is formatted and when information is needed from other stakeholders for decision-making. Ensure that everyone on the project team understands the information needs and expectations of one another to help map out information flow during the design process.

Spend time determining BIM roles and responsibilities, including how to best manage collaboration when using the model.

Other logistics with IT

Define and negotiate as a team what you need in an IT system for the project. Every project team member will have different comfort levels and skill sets with different technologies. Determine together as a team what technology you need to accomplish your project. Consider the time needed for project team members to transition and train in a new IT system. Then, negotiate among project team members what technologies to adopt for the duration of the project.

Make sure your information system is scalable, searchable, and anticipates project needs. Choose an information system for storing and finding information that is scalable to allow it to grow with the project. Anticipate what your project information needs will be at the end of the project when making your information-system selection.

Related chapters include: **Design Management** (pg. 73)

RISK MANAGEMENT
How Do We Track and Manage Risk?

What is a risk register?

There is no traditional "contingency" on an IPD project, instead IPD teams use a risk register (also called a risk log) – a detailed list of all potential issues on a project that are either a risk or an opportunity for the project, as related to budget, schedule, or other measurable outcomes. Opportunities track innovation and stretch goals. The risk register is typically a spreadsheet in which each row represents an item; a rough sense of its impact (at a minimum, the cost and schedule); the probability of it actually occurring; whether it is a priority, an IPD item, or a potential owner change; who is managing the item; and next steps.

Why is a risk register important?

A risk register keeps the project team focused on consistently identifying risks, uncertainties, and opportunities for technical and budget issues; planning for implementation; and improving certainty. It indicates what the risks of a project are and—depending on the nature of the risk—who is responsible for them. This is a key tool for knowing how to and who is managing risks on the project. It is not meant to track team culture. *(See Team Management for more information on those metrics.)*

An effective process for using the risk register is for the project team to proactively identify possible risks and uncertainties, discuss these as a team, make a plan to manage or mitigate the risk, and follow through and update as a team, with discipline. Place items on the risk register as soon as possible. Even an issue with

a 1% probability should go on the register to permit the project team to start to actively manage the risk. Aggregate the sum of the total risk and weighted risk to give the project team an idea of the projected finances, schedule, and other performance of the project early on. This allows the project team time to make adjustments in order to maintain progress toward their targets.

How do I create and use a risk register?

Start early. Have the project management team (PMT) develop a risk register during the early planning stages of a project.

Have team sessions to anticipate risks. Ensure that the PMT comes together in an initial working session to go through and list the potential risks of a project, and makes the review of that list a standing agenda item.

Determine categories of risk. Risk comes from many sources: budget, schedule, program, regulations, and personnel (i.e., the risks associated with the departure of key project team member or with new project team member). Determine what categories of risk need to go into the risk register.

Determine the value, impact, likelihood, and priority of each risk. During each team session, assign each risk a dollar value (even if it is only an order of magnitude or a range, and/or a schedule-impact value), as well as information regarding its potential impact, likelihood of occurrence, and its priority for resolution.

Assign a champion to each risk item. The champion of a risk will be responsible for organizing the project team members to help solve the problem and manage the risk.

Establish a plan to mitigate risks. After each session, have team champions organize project implementation teams (PIT) to create a plan to mitigate risks.

Assign a date to resolve each risk. Have the champion of each risk assign a due date in the risk register for when the team believes the risk can be resolved and retired from the register.

Tie the risk register to track-contingency utilization. On the assigned due date, have the champions report to the IPD team, whether a risk is resolved or requires additional contingency to the budget or schedule, or other actions. Tying together and tracking the contingency with the risk log shows what risks are taking away from the contingency and what risks are being resolved and replenishing the contingency.

Continue to update and amend the risk register as a team throughout the project. Throughout design and construction, continue to reformat and update the risk register based on the work and activity of PITs.

Risk-and-opportunity register option. Consider using a combined risk-and-opportunity register in your process to constantly stimulate multidimensional thinking and constant attention to improvement. *(See Appendix 10 for an example of a risk and opportunity log.)*

There is no traditional "contingency" on an IPD project. Instead IPD teams use a risk register—a detailed list of all potential issues on a project that are either a risk or an opportunity for the project.

Related chapters include: **Team Management** (pg. 41)

AN OWNER ON THEIR RISK REGISTER

One owner describes the log they use for weighting risks and opportunities using rough order of magnitude (ROM). "You have a risk (drainage issues for example), if it occurs, would cost $500k. Continuing with this example, you have soil boring information that shows it is unlikely to occur, perhaps it is 20% probability. In the weighted system, we assign a dollar amount to this item of the possible cost multiplied by the probability of its occurrence. For this example, the amount is $100k ($500k x 20%). Adding up the weighted costs for all risks and all the potential savings (opportunities) results in a number you compare to your available contingency. You update ROMs as you get more information about the probabilities. When an item goes to zero because it's so improbable or reaches 100% with a known cost, it comes off the log.

PROJECT DASHBOARDS
How Do We See If We Are On Track?

What are project dashboards?

High-functioning teams have lean visual management systems, sets of quantitative and qualitative metrics, and data tracking summarized in simple-to-read visual formats known as dashboards. Dashboards include tracking performance against budget and schedule, as well as other key performance indicators, such as safety, program, culture, profit, contingency, and usage.

Why are project dashboards important?

Dashboards are a practical and powerful way to track a project's progress. Tracking and measuring progress is key to identifying leading indicators in time to adjust and succeed, and to noting red flags that require prompt attention. Dashboards also make visible and apply team ownership to specific project issues to enable prompt decision-making. When comprehensive dashboards are reviewed by project leadership on a regular basis, the project team can adjust quickly as issues arise rather than being too late to fix them.

How do you decide what to measure?

Deciding what to measure is dependent on the unique goals of the project. What you choose to measure should help with decision-making and determining whether the project is a success or a failure. For example, monitoring and tracking key schedule milestones over time will give you an accurate picture of whether you are making good progress on the project and if the project team is effectively minimizing overtime hours.

To decide what to measure:

Make metrics a part of your early conversations about goals. Use this time to mesh owner Conditions of Satisfaction, project goals (e.g., project safety, lean, sustainability), and project team goals (e.g., firm goals, professional goals).

Identify metrics needed to measure the success or failure of goals. Once you identify goals, you can identify what metrics are needed—early and often—as leading indicators that will stimulate action to achieve project outcomes. Determine how you will measure and collect data.

Ask your team. Some experienced integrated teams may have suggestions for the types of metrics and tools that will benefit the project.

How do we measure progress?

Once you decide what to measure, make a list of the types of data you will need and then identify the tools you need to collect and measure the data. Be sure to:

Keep it simple. Choose the tool that fits your needs. Do not select a tool that is going to give you more information than you need. Ideally, information should be legible "at-a-glance".

Use a kit of tools to track multiple types of data. Use a combination of different tools for different types of tracking. Examples of tools in a project tool kit include budget-tracking software, regular online surveys, and Plus/Deltas.

Quantify soft skills. Assessing the health of your project team's culture is highly valuable on an IPD project. When measuring project team morale and performance, quantify team behaviors and values. Using a Likert scale (a type of rating scale) for responses can help your project team track member concerns and identify areas for improvement. Consider whether you want team member responses to be anonymous or identifiable. While an anonymous survey might make team members more likely to open up about concerns, having identifiable survey responses means you can identify where or with whom improvement or support is required. (For examples of how to measure team morale and performance, see Appendices 19, 20, and 21.)

How do I make the most of project dashboards?

Use visual control. Visual control makes the data explicit, standardized, and quick and simple to understand. When a dashboard uses effective and meaningful visuals, you can easily see how big or small a problem is on the project. As one contractor noted, "You cannot manage what you cannot see." Simple, but effective,

dashboard visuals might be things like a thermometer divided into red and green zones or a spider graph with axes of important criteria. *(See Appendix 20 for example of a spider graph.)*

Avoid complexity. Choose measures and tools that make sense for the project and project team. If collecting and measuring data becomes too complicated, your team members will spend more time collecting and managing data and less time getting their actual work completed.

Consider an online global dashboard. These dashboards, such as iDashboards, Tableau, Smartsheet and others, use a web interface for data visualization and allow project team members to plug in their data using a global database in the cloud, helping your project team generate real-time data for the project. They also allow project leaders to access the dashboard and database at any time. Since web tools can often remain hidden if project team members are not checking them regularly, make this dashboard visible to project team members by displaying it in a hallway, co-location space, or conference room. *(See Appendix 14 for examples of dashboards in a Big Room.)*

BENCHMARKING AND JUSTIFYING GOALS TO TRUSTEES

Collecting and measuring project data is not only important for tracking a project's progress but also allows you to benchmark your IPD projects. For example, you can measure the difference between project financials, such as your IPD project's conceptual project estimates and final estimates, your burn rate through project and team performance contingency, or change-order rates. You can then compare these to past and future IPD or non-IPD projects to determine whether your organization, or IPD projects, is improving.

Metrics can also be used to justify the goal of the project to your board of trustees. Showing a board of trustees your intended metrics for a project can help them understand project goals and how the project team will track those goals.

Financial Metrics	What percent of the budget is secured (e.g., bought out, committed) over time
	Warranty period tracking (especially if there is a performance incentive)
	Cost Slippage
	Tracing rework (e.g. type of rework, defects)
Schedule Metrics	Schedule slippage
	Schedule-milestones tracking
	Change-order time
	Earned Value Analysis
Productivity Rates	Productivity rates for trade contractors (e.g. budget, linear feet of material per man-day)
	Productivity rates for design (e.g. deliverable commitments, decision-making)
Safety	Number of safety recommendations
	Number of safety incidents
	Number of man hours
Percent Plan Complete	PPC on design
	PPC on construction
	PPC on management in the field
Value Achievement	Value adds, including BIM value adds
	Change-order rate
	Contingency burn rate
	Quality "done right" versus percentage of rework
	Punch list items
	Quality conformance and nonconformance
	Issue resolution efficiency
Team Morale and Team Performance	Team values (e.g. health-check on team alignment with goals)
	Innovation and good-ideas tracking
	Behavioral values
	Accountability
	Team accomplishments

Dashboards are a practical and powerful way to track a project's progress. Tracking and measuring progress is key to identifying leading indicators in time to adjust and succeed, and to noting red flags that require prompt attention.

Dashboards also make visible and apply team ownership to specific project issues to enable prompt decision-making.

Use portfolio-management tools across projects. From the start of your first IPD project, you will want to track variances between your cost estimates and the actual cost of the project. With these metrics, you can use portfolio-management tools to track across projects to see if you are improving at budgeting IPD projects.

Implement and manage metrics in design and construction. Begin forecasting early. Metrics will show what has changed and determine actions for making improvements.

Identify a metrics champion. Have someone lead the effort for specific data collection and management.

Review dashboards regularly with your team. Regular team review will help to identify potential team and project issues early, providing plenty of opportunities for developing countermeasures to solve an issue. When reviewing dashboards, have them visible to all project team members.

Have conversations around countermeasures. When your dashboard identifies a problem on the project, have a conversation around the problem and suggest three or four countermeasures that might solve the issue. These measures should include specific actions and time frames. After implementing one or more of these countermeasures, check in with the project team to see if the actions are working. (*See Appendix 21 for an example of how metrics led to team improvement through establishing countermeasures.*)

DEVELOPING SKILLS FOR VISUAL CONTROL

Using dashboards effectively requires a visual skill set that not all team members have. As one contractor commented, "The admin on the project needs to have different skills to create the graphs and the charts....It takes time to create visual controls."

Modify the dashboard as needed. Using dashboards is an iterative process. If the dashboard does not identify a problem before it occurs, modify how you collect, measure, and visualize your data.

Maintain tracking throughout the project. One of the biggest challenges for most project teams is to continue tracking throughout the project. Ensure that all project team members understand the value of tracking metrics and how those metrics will help them improve their team and their project.

Celebrate success. When your dashboard shows that your project team has met a target or goal, celebrate your project team's success. This will help your team feel proud of their work and rally around the project.

Types of metrics

Dashboard metrics should include a project's finances, key schedule milestones, safety incidents, productivity rates, percent plan complete (PPC), value achievement, team morale, and team performance. Measuring across these metrics will help you identify challenges and solutions early in the project.

Deciding what to measure is dependent on the unique goals of the project. What you choose to measure should help with decision-making and determining whether the project is a success or a failure.

KEEPING IT SIMPLE WITH MANUAL DATA COLLECTION

It is important to keep it simple when tracking variances and PPC. On one project, a contractor would uses stickies and colored dots from the Last Planner System to identify the PPC and enter the percentage manually into the dashboard's database.

**PATH TO
CONTRACT**

Owner Alignment

Team Selection

The Contract Workshop/Team Alignment

**ONGOING
CONSIDERATIONS**

Team Management Building and Managing a Successful IPD Team

Financial Organization and Financial Monitoring

Lean Thinking

EARLY WORK

Validation: Go/No Go

Target Value Design

Co-location in a Big Room

Design Management

Prefabrication

Integrating Project Information Using Building Information Modeling

Risk Management

Project Dashboards

LATER WORK

Team Maintenance

**WHAT GOES
WRONG**

Closeout

LATER WORK
Maintaining Momentum

If you take the time needed to build a strong foundation for communication, collaboration, and accountability during your early IPD work, much of the later work during construction and closeout will consist of maintaining team and project momentum and adjusting as things evolve. You need to ensure that your tools and techniques are sustained and working effectively. You also need to make sure that your team culture preserves its strength and that processes for managing the work and maintaining financial diligence continue to be carried out as planned.

Maintaining project team member continuity is critical for the ongoing success of a project as team members who remain on the project have a wealth of knowledge about the project's history, goals, and drivers for decision-making. To have team continuity, you need to maintain a strong team culture that allows team members to continue to develop their skills. Strong team culture fosters individual leadership and personal development, defines clear roles and responsibilities, and provides continuous training. Team continuity also requires that there are mechanisms in place when a team member leaves to ensure a smooth transition (off-boarding), as well as mechanisms to bring new team members up to speed on the project, processes, and culture of the team (on-boarding). On-boarding and off-boarding are an essential part of forming and maintaining a strong team culture.

This section ends with a discussion on what to expect and do during a closeout. This includes planning for the desired deliverables needed for this period, as well as finalizing the financial terms of the project.

TEAM MAINTENANCE
What Do We Do When People Come, Go, or Become Complacent?

To keep teams efficient and effective, put into place processes and mechanisms that provide a smooth transition as new members join the team—such as formal on-boarding steps or ground rules for behavior. Secure team members on the team and ensure that they stay committed to the project. If a key member leaves, make sure off-boarding ensures a smooth transition. All project participants contribute to team culture, however IPD team members have a special role to lead and model collaborative behavior supporting innovation. All project team members should experience some level of the processes described below.

All project participants contribute to team culture, however IPD team members have a special role to lead and model collaborative behavior supporting innovation.

On-boarding

What is on-boarding?

On-boarding is the process used to continually align team members and get new team members up to speed on a project. On-boarding aligns new team members to owner and project goals, defines expected collaborative team behaviors, and explains expected work processes. Having a robust on-boarding process for team members will help to manage design and construction, maintain team continuity, and build a resilient team that can handle and resolve project challenges efficiently and effectively as they arise. On-boarding is important to all project team members, IPD team members may need a more in-depth on-boarding process.

Why is on-boarding important?

Introducing new members to the project team can be challenging. The IPD team has already spent many hours together in early planning, aligning their goals, and identifying the many large and small shifts in their behaviors and attitudes required to move from a traditional to an IPD team. New members need to be caught up quickly while still giving them opportunity to shape the team as it evolves. New members do not need to know every detail of what happened before they joined, but they must fundamentally understand that IPD is different from traditional forms of project delivery. For project team members not on the IPD team, basic understanding is very important. The primary goal of on-boarding is for all project participants to know that even if their company's profit is not at stake, collaborative behavior is expected.

Who conducts on-boarding?

On-boarding can be conducted by the IPD team, owners, or a third-party coach that specializes in IPD collaboration. For craft labor, teams have found it effective for on-boarding to be done by the project "last planners," typically superintendents and general foremen.

How do I start on-boarding team members?

Start the on-boarding process early. On-boarding is an investment that requires a high level of time and personal commitment for owners and project team members. Start on-boarding early in the project, during the validation process.

Start with the IPD team. On-boarding should begin with the essential IPD team members from each of the signatory companies: typically the general contractor, architect, and trade partners. Early on-boarding of trade partners guarantees that they can collaborate in the design process. You can continue on-boarding other IPD team members and project participants as they join the project.

How do I get a new team member up to speed quickly?

Educate the team member about project goals. Prior to on-boarding, all project goals should be documented in the project charter and Conditions of Satisfaction (CoS). As every project decision will need to be filtered through the lens of the project's primary goals, educate team members about the goals and mission of the project. *(See Appendices 2 and 4 for examples of CoS and project values.)*

Educate team members about a project's CoS. The CoS provides the team with agreed-upon criteria for decision-making that will achieve the project's goals. This will help align the team around project goals and build shared understanding around how decisions will be made on the project.

Consider repeating selected training. If collaboration-skills workshops or other training were helpful early in the project, consider repeating for those team members who joined late and did not participate in the original training. IPD team members joining after the project begins may need training on the financial and contract terms.

EDUCATE ALL TRADE CONTRACTORS ABOUT THE CONDITIONS OF SATISFACTION (COS)

Educating new project team members about the project's CoS provides a foundation of knowledge that can help project team members understand decision-making drivers. As one expert noted, "Say you are a plumber and you come onto the floor and all the other trade contractors are educated about the CoS for this floor. For example, they know this is the ICU; it is going to be state of the art; it is going to do these things; and it is going to serve these people. That can be built into on-boarding. For the person doing the work, that knowledge base is really important."

What else should I do with new team members?

Share the established Big Room management rules. Ensure that the project team is well versed in Big Room management. Team members should learn how to set up and run a Big Room that acts as a highly integrated, collaborative space for the team. Some teams have used buddy systems—new team members are assigned an experienced mentor for the first few days.

It takes time to bring new team members up to speed on team rules, work processes, and how to engage in a highly collaborative setting. Give your participants time to learn how to collaborate effectively and efficiently.

Share the agreed-upon strategy and tools for documenting and communicating decisions. Ensure that all project team members understand the decision-making process—that they understand how to document, communicate, and track decisions.

Communicate the agreed-upon ground rules on team behavior. Having a team that exhibits positive, collaborative behaviors is key for team resilience. Ground rules on behavior can include not only how to communicate and interact with one another but what are the protocols when deliveries arrive at the job site. Types of team behavior ground rules that should be established during on-boarding include:

> Facilitation behaviors;

> Leading by example;

> Honoring every opinion; and

> Tiered communication structures.

Give people time to get up to speed and engage. There is a steep learning curve for participants on IPD projects. It takes time to bring new team members up to speed on team rules, work processes, and how to engage in a highly collaborative setting. Give your participants time to learn how to collaborate effectively and efficiently, particularly the IPD team members.

BE RIGID AND DISCIPLINED ABOUT HONORING GROUND RULES

If somebody is not following a ground rule, that is an immediate warning that they are not behaving the right way. If this occurs, figure out why someone is not following these rules and work toward a solution to return to positive team behavior.

JUSTIFY THE VALUE OF TOOLS AND PROCESSES TO NEW TEAM MEMBERS

Whether using lean tools, like the LPS or BIM, make sure to demonstrate the value of these tools to the team members who are new to IPD. For example, if there is a new trade contractor that has never used LPS, offer a value statement as to why it is important to use LPS with the group.

Team Maintenance and Team Member Continuity

What is team maintenance and team continuity?

Team maintenance is the work needed to ensure that the established team culture evolves and remains strong. This includes the IPD team and the larger project team. As the project progresses from early to middle to late stages, the team adapts to address different challenges, often with different personnel in leadership roles. Additionally, over time, complacency can creep and inertia may pull some members back to traditional roles and behaviors. It is important to reinforce IPD culture on a regular basis to maintain energy and momentum of the team. Strong team maintenance helps safeguard team continuity: that team members—and the wealth of knowledge that they have about the project's history—remain on the project.

Why is team maintenance and continuity important?

Maintaining team member continuity is important for the success of a project. You can set up strong team culture, but the team needs to continually support leadership, evolve its structure, and provide timely training. Adjustments may be needed as the project evolves and individuals come and go. IPD projects have intense team interaction, sometimes over long periods of time. In projects of long durations, teams can lose momentum or find that processes that worked in the beginning of the project are no longer effective later.

How do I continuously engage team members?

Providing new opportunities for individuals, encouraging innovation, utilizing tracking and metrics, encouraging positive team member behaviors, and ongoing check-ins with the team all contribute to a strong team culture.

Encourage different leadership and provide growth opportunities. Large IPD projects are carried out over a long period of time. This means that team members can get stuck within the same scope of work on a project. Make sure different team members from different disciplines have a chance to lead during the project and challenge themselves with new opportunities.

TIERED COMMUNICATION STRUCTURES

Having ground rules around tiered communication structures sets expectations about how and to whom certain issues should be addressed and when issues are deferred to higher-level decision-makers in the team structure. Setting these expectations can help build a sense of emotional safety on your team and help maintain trust.

Over time, complacency can creep and inertia may pull some members back to traditional roles and behaviors. It is important to reinforce IPD culture on a regular basis to maintain energy and momentum of the team.

How do I ensure smooth transitions if someone leaves?

Track innovation and ideas. To maintain a culture of innovation and open communication, track innovative and valuable ideas from team members during the project. Invite ideas from everyone, including trade contractors. Recognize good ideas that are brought to the team and track them over time to show how the team has grown on the project.

On-board new project participants. While on-boarding often occurs during design, construction has not traditionally conducted continuous on-boarding of new team members. Yet, there is a huge learning curve for construction workers around IPD processes and culture. When bringing in new team members, such as a foreman, superintendent, or laborer, on-board them to align them with the goals and expected team behaviors of a project. *(See On-boarding for more information on setting expectations for team behaviors.)*

How do I continue with coaching and training?

Coaching helps improve the team through repetition of good behaviors and work practices. This support is typically focused on IPD team members but may include other participants. Not all IPD projects require the use of a coach or facilitator, but they are useful if this is you or your team member's first IPD project. *(See Team Management for more in-depth discussion of facilitation.)* To continue with coaching and training:

TO MAINTAIN A STRONG TEAM CULTURE, REMEMBER TO:

> Continue monitoring team culture and behaviors throughout the project, using metrics if appropriate;

> Continue engaging in and encouraging positive behaviors, for example, focus on being tough on process but easy on people;

> Continue celebrating success;

> Continue using positive meeting structures for engagement; and

> Make it possible to be challenged.

If collaboration-skills workshops or other training were helpful early in the project, consider repeating for those team members who joined late and did not participate in the original training. IPD team members joining after the project begins may need training on the financial and contract terms.

Bring in a coach or facilitator to reinforce collaborative team behaviors. When things go wrong on a project, the team may need further training and coaching to reinforce positive behaviors. During the project, look for signs that your team needs more coaching or facilitation.

Identify future facilitators on the team.

Have a coach or facilitator teach facilitation skills to team members. Good facilitation skills can come from any team member regardless of their traditional role. Look for naturally strong facilitation skills among your team members or owner representatives and encourage them with training and feedback.

As the facilitator phases out, make sure that team leaders become stewards for collaboration.

While coaches and facilitators can help your project team learn and develop stronger collaborative behaviors, make sure that the IPD team is still seen as the leaders in the room. Identify team members, such as project managers or another team leader, who have the capacity for facilitating. This way, when the facilitator phases out of the project, a team member can step into the facilitation role.

Related chapters include: **On-boarding** (pg. 93), **Team Management** (pg. 41)

STICKING UP FOR THE TEAM

One owner described a time where defending his team members increased trust. The team had already been selected, and an owner-executive, who had not participated in the selection process, rejected the choice of the engineering firm. When the owner asked the team their thoughts, the architect said, "You told me to bring a team. I brought a team. If you're going to get rid of the engineers, I'm walking out the door with them." As the owner recalled, "That was the first time I'd ever seen a designer stick up for an engineering consultant. I went back to my boss and I said, 'I want to keep the engineer on the team.' And he said, 'The outcome is on you.' The construction manager told me afterward: 'You have no idea how much political capital you bought among the trade partners when you didn't kick the engineers off. When you defended the team, they knew they could trust you.'"

Off-boarding

What is off-boarding?

Off-boarding is the process of removing someone from the team, or "voting them off the island."

Why is off-boarding important?

Off-boarding processes should be well established on the team to make the consequences of poor performance clear to nonperforming team members. Typically focused on IPD team members, it can also include other project participants. It also offers them a chance to correct their performance. (*See What Goes Wrong—And What Can We Do About It regarding teams who regret delaying the off-boarding decision, explained in the Not Cutting Losses section.*)

Who does off-boarding?

It depends. Who determines a participant's removal should be established early in the project. Owners, the senior management team (SMT), or the project management team (PMT) can take the lead on off-boarding team members who are not performing. The contractor is often responsible with off-boarding trade contractors.

How do you plan for problems?

Before there is a problem:

Define and communicate criteria for removal. Your project should begin with well-defined criteria for removing a company or individual from the project teaml. The most common criterion is the evaluation by the team that one party is placing their own success above the project and team. Circulate these criteria and make sure that the team understands them.

Define and communicate the process and rules for removing team members. Early in the project, the IPD team should understand the process and rules for removing team members if they do not adhere to the rules or continuously do not meet commitments and expectations. Have IPD team leaders (e.g., the SMT and/or owner) define the rules and process for removal early and communicate them to the whole project team.

How do you prevent problems?

Have the PMT evaluate the team at regular intervals. To maintain a healthy team, have the PMT evaluate their team on a regular basis (e.g., weekly, monthly, quarterly). Use metrics to make evaluations. By conducting a regular review, you will be able to catch problems early and respond.

DESIGN MEETING STRUCTURE TO PROMOTE ENGAGEMENT

If you find team members are not prepared for meetings, change the meeting structure. For example, on one IPD project when several subtrade foremen arrived to meetings without preparing information, the superintendant assigned each of them a turn at running meetings, and their overall engagement improved.

Raise issues early. Do not hesitate to bring up a problem with an individual or firm working on the project. Waiting will just allow the problem to fester and will not help the team or your project.

Determine whether the issue is a company or individual problem. Sometimes difficulty is due to a team member not adhering to ground rules or behaviors. Other times, the individual may not have their own firm's support and may have needed resources on the project. Determine where the issue lies to correctly direct your concerns.

Send guidance on how to correct the issue. The PMT should send guidance to the individual on how to correct the problem. Assess whether the person is receptive to the guidance and attempts to make corrections to solve the problem.

How do you remove someone from the team?

Present the case for removing a team member by referencing the agreed-upon criteria. Have a meeting in the Big Room to present the case for removing someone or a company from the team. If it is part of your process, the team member or company can also present their own defense for staying on the project and how they will rectify the problem.

Follow your process. After the case is made for removal, follow your process to vote or otherwise decide on removal.

How do I deal with the aftermath?

Off-boarding can incite a range of responses, some of which can be contradictory. It can provide immediate relief to consistent tension in a team, raising morale. It can also be an affirmation to the rest of the team that their criteria and processes were fair and working as intended to keep the project on track. However, it may also lower morale if the process was perceived to be unfair, creating a sense of dread that anyone could be removed next. Having a candid post-off-boarding discussion can help strengthen team culture.

Related chapters include: **What Goes Wrong—And What Can We Do About It** (pg. 105)

TRACKING GOOD IDEAS

"One of the things that we have done is list who comes up with a good idea. It takes time. At first you will get [joke] suggestions from Mickey Mouse and Superman because they do not believe you are serious. We offered a $50 dollar gift card to Home Depot if somebody brought up a good idea that we used. It was not about who won. Bringing somebody forward during the superintendents and foremen meeting and saying this was Bob's idea and it's what we are going to do—that was more meaningful than anything else. After people were recognized, we started getting good ideas come out of the woodwork."

CLOSEOUT
How Do We Put a Bow on It?

What is closeout?

Closing out an IPD contract can take more effort than a traditional lump sum or guaranteed maximum price contract due to the cost-reimbursed agreement. All costs must be received to calculate the final cost. Once the final cost is known, the IPD team can determine if an incentive is due for shared savings, if all the profit is earned, or if the project overran the budget. Final invoices are paid, and the contract can be completed.

Why is closeout important?

It is important to close out the project once the owner puts the project into service for its intended use. The IPD team will want their portion of the profit that has been withheld (if due) and any earned incentive. The owner will want to stop the cost-reimbursed nature of the contract and start the warranty period.

How do you close out an IPD project?

Start the project with the end in mind. Map out the process for each required deliverable due at the end of the project with the IPD team and the owner's team. This includes but is not limited to:

➤ As-builts: discuss and agree on types of information and format (e.g., pdfs, models);

➤ Warranties;

➤ Lien Releases; and

➤ Detailed Invoice backup for capitalization.

It is important for the IPD team to manage the punch list to be able to complete it as soon as possible after the project is considered substantially complete. A punch list that runs long after a project is in service can result in significant additional cost to the owner. A punch list poorly managed by the team can also lead to a reduction of savings—and potentially profit.

Other things to remember are:

Have retrospectives during the warrantee period. Hold a retrospective with your project team during the warrantee period to see what went well and what could be improved. Share the lessons learned with your team.

Start milestone activation during design. Schedules may change, but the start and end dates do not. Determine and estimate the timing of activation milestones at the beginning of project design to provide your IPD team with a buffer during the activation period. This will help your team accurately predict what you need to do to get to the end date and open the building on time.

Set a deadline for warrantees. After a project is complete and turned over for its intended use, costs can continue to go up for warrantee items. It is important to draw a firm line at the end of a project when the cost-reimbursed nature ends. Any costs for warrantee work after that period will not be a direct cost to the project. It takes time to reconcile all final costs at the end of the project to calculate the incentives. Consider stopping direct billing to the project as quickly as possible after a punch list period.

PRE-VALIDATION
VALIDATION
DETAILED DESIGN AND IMPLEMENTATION DOCUMENTS

PATH TO CONTRACT

Owner Alignment

Team Selection

The Contract Workshop/Team Alignment

ONGOING CONSIDERATIONS

Team Management Building and Managing a Successful IPD Team

Financial Organization and Financial Monitoring

Lean Thinking

EARLY WORK

Validation: Go/No Go

Target Value Design

Co-location in a Big Room

Design Management

Prefabrication

Integrating Project Information Using Building Information Modeling

Risk Management

Project Dashboards

LATER WORK

Team Maintenance

WHAT GOES WRONG

Closeout

WHAT GOES WRONG—AND WHAT CAN WE DO ABOUT IT

―

What can go wrong?

Although IPD is a reliable method for delivering projects, it is executed by real people in an imperfect world. Whether you are undertaking your first IPD project, or your tenth, problems will occur. When they do, a rapid team response is, ideally, spurred by the futility of blaming others coupled with the common interest in quickly solving problems to preserve the profit pool. In addition, the IPD team forges relationships that will allow the participants to weather adversity. IPD is resilient, but it is unrealistic to believe that problems will not happen. Common problems that occur on IPD projects include:

Insufficient front-end planning, training, and coaching. Project teams get very nervous if they are not producing "real work." As a result, they often want to begin work before they really understand what they must do, the sequence in which it should be done, and the techniques they should employ. In addition, cash flow in IPD projects is front loaded, with more being spent earlier than in other project types. At the beginning, it may seem that a lot is being spent but not a lot accomplished. Because projects with aggressive financial targets are initially over budget, the pressure to develop cost reductions may seem at odds with investing resources on preparation and coaching. However, there is a real risk that avoiding front-end investment will result in more significant waste later. Project teams sometimes report that front-end planning could have been more efficiently managed, but most project teams feel that they would have benefitted from more front-end effort.

Insufficient/inefficient co-location. Some project teams have struggled with co-location. This may be due to project participants being engaged after key decisions were already made, project team members not wanting to spend time in co-location, or that the co-location was not properly organized and managed. Project teams that abandon co-location risk losing innovation and collaboration.

Reversion. Executing work in a new way is awkward and undermines feelings of competence. One project manager remarked that if it feels comfortable, you are doing it wrong. Virtually every IPD team has experienced project team members reverting to their traditional processes and behaviors. Yet, succumbing to reversion lessens performance. Faced with intense schedule pressure, project team members may think they do not have the time to learn a new skill—even if it will save time once it is mastered. They also may abandon skills that they have learned, for example, Last Planner System, as they near completion and schedule pressure increases. Great IPD teams stay focused on executing the project using lean and IPD principles, even when stressed by cost and schedule.

Unrealistic expectations. A great many IPD teams have achieved extraordinary results. Unfortunately, we have seen some parties assume that just executing an IPD agreement will result in miraculous results. Ironically, these parties are often those who do not see any need to change their own behavior. Unrealistic expectations paired with an unwillingness to change can only lead to disappointment, which leads to disenchantment, and then to abandonment.

Well-managed IPD teams carefully and continuously forecast costs and manage change promptly.

Inadequate cost/change management. In most IPD compensation models, the owner guarantees cost but not profit. Because the owner pays costs as they are incurred, some IPD teams have put change management below other project priorities. However, if an IPD team does not keep its risk, opportunity, and change logs current, it cannot know where it is financially and is not managing to target cost.

Not cutting losses. Sometimes people or companies just do not work out, and off-boarding is needed. Ideally, compatibility issues will be resolved during the selection process. However, in the real world problems arise after the project has commenced. This may be because of changed circumstances or just an incompatibility with collaborative work. If you do not cut your losses, there is significant research and practical experience to back up what every grandmother knows: "A bad apple spoils the barrel."

Company hubris. Companies enter into IPD contracts, but IPD projects are executed by people. There are several examples of good companies that have had very successful projects that then seem to assume that the knowledge gained on one project would automatically translate to a subsequent IPD team. This leads to waste as the new IPD team struggles to learn for itself what the earlier team has already learned. In a similar way, companies will transfer new personnel into an IPD project without putting them through the same level of training and on-boarding that the original IPD team received and just assume they will understand the process. They will not.

What can we do about it?

The University of Minnesota/IPD Alliance study reported that 90% of the combined participants (owners/designers/builders) believed that their IPD projects outperformed other forms of project delivery—a truly exceptional result.[7] However, the success of IPD projects does not mean that teams should be complacent. To keep improving, have teams focus on identifying and eliminating behaviors and practices that limit project success, and build on the behaviors and practices that are the core of IPD.

Use these strategies and solutions to prevent or correct the following problems:

Insufficient front-end planning, training, and coaching. It takes time and effort for teams to learn new ways of working, and lessons need to be repeated until they are fully understood. Training is a continuous process, not a singular event. Commit time and resources in front-end planning, training, and coaching before the benefits of IPD materialize. While it may be possible to do too much training, and it is possible to be inefficient in executing training, teams have never reported that they should have done less—often teams report that they should have done more.

Insufficient/inefficient co-location. Co-location is a powerful engine of collaboration. In addition to the recommendations in this guide and elsewhere, visit co-location sites to see what other teams are doing, engage co-location coaches, and develop a robust co-location management plan. (For more detailed recommendations, see Chapter 14 "Co-locating to Improve Performance" in *Integrating Project Delivery*.[8])

7   R. Cheng, M. Allison, C. Dossick, and C. Monson. *IPD: Performance, Expectations, and Future Use* (Minneapolis, MN: University of Minnesota and the Integrated Project Delivery Alliance, 2015).

8   M. Fischer, H. Ashcraft, D. Reed, A. Khanzode, Integrating Project Delivery (Hoboken, NJ: Wiley, 2017).

Reversion. When project team members begin to revert to old ways of working, use the following strategies:

> **Bring in external trainers and coaches.**
>
> Teams, especially project teams that are undertaking their first IPD project, benefit from external coaching and training. Coaching and training greatly accelerates team integration. In addition, it is often difficult to see your own behavior. An external coach—i.e., one that is not part of the project team—will be able to see dysfunctions unapparent to the team members.

> **Provide continual reinforcement.**
>
> Patrick Lencioni, in *The Advantage*,[9] counsels that team management must communicate their vision, and then overcommunicate. Have teams reinforce their project vision and expected communication and interaction behaviors with signs, pictures, charters, meeting rules, and other reminders.

> **Ask, what are we doing differently?**
>
> Have your IPD team periodically ask themselves what they are doing that differs from their prior projects. If the answer is "nothing" or "very little," then the team is on the wrong path. This is a good question to be on every project management team/core group agenda.

> **Empower everyone to declare "breakdown."**
>
> In a manufacturing setting, lean team participants are required to immediately notify management when a defect is discovered to allow corrective action to rapidly take place. In a similar way, IPD teams should immediately notify project management when the collaborative system is breaking down. Management should investigate, then develop and institute a corrective action plan. Demonstrate respect for team members who declare the breakdown for helping to preserve the collaborative environment.

Unrealistic expectations. Validation is the primary countermeasure for managing unrealistic expectations. *(See Validation for more information.)* It is appropriate, and even desirable, to set challenging goals. However, ensure that goals are grounded in a realistic understanding of the task.

Inadequate cost/change management. Collaboration requires accountability. Well-managed IPD teams carefully and continuously forecast costs and manage change promptly. Summarize this information on a near-real-time basis and display the information in well-designed dashboards that provide the information necessary for decision-making. Display these dashboards in the Big Room to allow everyone to know where the project is, where the project needs to be, and to focus on getting it done.

Not cutting losses. If a problem arises, the first step is to try to counsel, train, and support. At some point, however, the IPD team will have to decide whether further efforts to support a team member or company are worthwhile or whether a change should occur using their off-boarding process. While it is difficult to terminate an IPD company or to request that project participants be changed, when that has occurred the IPD teams generally reported that they should have made the change earlier. Have an off-ramp for problematic people or organizations in your IPD agreement.

Company hubris. Many companies have recognized that they needed to take positive steps to assure that learning on one project carries over to the next. This can be accomplished by using internal IPD champions who provide training and continuity across projects, some of the same personnel who can bridge between projects, or the same external IPD coach between projects who can carry over lessons learned.

---

9 P. Lencioni, *The Advantage: Why Organizational Health Trumps Everything Else in Business* (San Francisco, CA: Jossey-Bates, 2012).

Related chapters include: <u>Validation</u> (pg. 63)

MORE RESOURCES

American Institute of Architects (AIA), AIA Minnesota Council, and the School of Architecture, University of Minnesota. *IPD Case Studies*. Washington, DC: American Institute of Architects and AIA Minnesota, 2012. http://z.umn.edu/ipdcasestudies2012cheng.

American Institute of Architects California Council (AIA CC). *Integrated Project Delivery: An Updated Working Definition*. Version 3. Washington, DC: American Institute of Architects and AIA California Council, 2014. http://leanipd.com/wp-content/uploads/2017/11/IPD-A-Working-Definition-FINAL.pdf and http://www.aiacc.org/wp-content/uploads/2014/07/AIACC_IPD.pdf.

American Institute of Architects and AIA California Council (AIA/AIA CC). *Integrated Project Delivery: A Guide*. Version 1. Washington, DC: American Institute of Architects and AIA California Council, 2007. http://www.aia.org/contractdocs/AIAS077630.

Ashcraft, Howard. "Integrated Project Delivery." In *Construction Law Handbook* (2018 supplement), edited by Stanley A. Martin and L. Rochwarg. New York: Wolters Kluwer, 2018.

Ashcraft, Howard. *The IPD Framework*. San Francisco, CA: Hanson Bridgett, 2012. https://www.hansonbridgett.com/-/media/Files/Publications/IPD_Framework.pdf.

Ashcraft, Howard. *Negotiating an IPD Agreement*. San Francisco: Hanson Bridgett, 2010. https://www.hansonbridgett.com/-/media/Files/Publications/NegotiatingIntegratedProjectDeliveryAgreement.pdf.

BIMForum. "Level of Development Specification." http://bimforum.org/lod/.

Center for Innovation in the Design and Construction Industry. http://www.CIDCI.org.

Cheng, Renée. *Motivation and Means: How and Why IPD and Lean Lead to Success*. Minneapolis: Integrated Project Delivery Alliance (IPDA) and Lean Construction Institute (LCI), 2016. http://arch.design.umn.edu/directory/chengr/documents/motivation_means2016.pdf.

Cheng, Renée. *Integration at Its Finest: Success in High-performance Building Design and Project Delivery in the Federal Sector*. Washington, DC: Office of Federal High-Performance Green Buildings, U.S. General Services Administration, 2015. http://arch.design.umn.edu/directory/chengr/Integration_at_its_finest.pdf.

Cheng, Renée. *Teams Matter: Lessons from ARRA*. Chicago: Region 5, U.S. General Services Administration, 2016. http://arch.design.umn.edu/directory/chengr/documents/TeamsMatter_RCheng.pdf.

Cheng, Renée, Markku Allison, Carrie Sturts Dossick, and Christopher Monson. *IPD: Performance, Expectations, and Future Use*. Minneapolis: University of Minnesota and the Integrated Project Delivery Alliance, 2015. https://ipda.ca/site/assets/files/1144/20150925-ipda-ipd-survey-report.pdf.

Cohen, Jonathan. *Integrated Project Delivery*: Case Studies. Sacramento, CA: AIA California IPD Steering Committee, AIA IPD Group, 2010. https://www.ipda.ca/site/assets/files/1111/aia-2010-ipd-case-studies.pdf.

Computer Integrated Construction Research Program. *BIM Project Execution Planning Guide—Version 2.1*. University Park: Pennsylvania State University, 2011. http://bim.psu.edu/.

ConsensusDocs. https://www.consensusdocs.org/Catalog/collaborative.

Dal Gallo, Lisa, Shawn T. O'Leary, and Laila Jadelrab Louridas. *Comparison of Integrated Project Delivery Agreements*. San Francisco: Hanson Bridgett, 2010. https://www.hansonbridgett.com/-/media/Files/Publications/IPD_Contract_Comparison.pdf.

Fischer, Martin, Howard Ashcraft, Dean Reed, and Atul Khanzode. *Integrating Project Delivery*. Hoboken, NJ: Wiley, 2017.

Integrated Project Delivery Alliance (IPDA). https://www.ipda.ca/.

Lean Construction Blog. http://leanconstructionblog.com/.

Lean Construction Institute (LCI). https://www.leanconstruction.org/.

Lean Construction Institute—Canada (LCI-C). http://www.lcicanada.ca.

Lean IPD. https://leanipd.com.

Leicht, Robert, Keith Molenaar, John Messner, Bryan Franz, and Behzad Esmaelli. *Maximizing Success in Integrated Projects: An Owner's Guide*. Version 1.0. University Park, PA: Pennsylvania State University, 2015. Available at http://bim.psu.edu/delivery.

Lencioni, Patrick. *The Advantage: Why Organizational Health Trumps Everything Else in Business*. San Francisco: Jossey-Bates, 2012.

Liker, Jeffrey K. *The Toyota Way: 14 Management Principles from the World's Greatest Manufacturer*. New York: McGraw-Hill, 2004.

National Association of State Facilities Administrators (NASFA), Construction Owners Association of America (COAA), APPA: The Association of Higher Education Facilities Officers, Associated General Contractors of America (AGC), and American Institute of Architects (AIA). *Integrated Project Delivery: For Public and Private Owners*. Washington, DC: American Institute of Architects, 2010. https://www.coaa.org/Documents/Owner-Resources/Industry-Resources/IPD-for-Public-and-Private-Owners.aspx.

"Off-Site Construction Council: About the Council." *National Institute of Building Sciences*. Accessed 3/2/2018. http://www.nibs.org/?page=oscc.

PS2L Project Production Systems Library. "Target Value Design (TVD)." http://p2sl.berkeley.edu/research/initiatives/target-value-design/.

Thomsen, Chuck, Joel Darrington, Dennis Dunne, and Will Lichtig. *Managing Integrated Project Delivery*. Report for Construction Management Association of America CMAA, 2010.

Tommelein, Iris D. and Glenn Ballard. *Target Value Design: Introduction, Framework & Current Benchmark*. Arlington: Lean Construction Institute, 2016.

Umstot, David and Dan Fauchier. *Lean Project Delivery: Building Championship Project Teams*. Self-published, CreateSpace, 2017.

Williams, Joan, Su Li, Roberta Rincon, and Peter Finn. *Climate Control: Gender and Racial Bias in Engineering*. San Francisco: Center for Worklife Law & Society of Women Engineers, 2016.

GLOSSARY

A3

A one-page report on a single 11 x 17 sheet of paper, which uses PDCA thinking as it applies to collaborative problem solving, strategy development, or reporting. An A3 includes a problem statement, data and background information, analysis, proposed options, recommendations and agreements, actions, expected results, and follow-through. (*See Appendix 12 for an example of an A3.*)

A3 Thinking

A3 Thinking refers to the structured process of documenting a problem, solution, and action plan. The A3 Thinking process is undertaken collaboratively, with input from all stakeholders on the topic. It begins with consensus on the problem statement and arrives at consensus on a solution and path forward.

Actual Cost

The sum of the total cost of the work actually incurred by the project participants in connection with the performance of all phases of the project. Does not include owner expenses, such as fees for permit, inspection, or equipment. Depending on the contract form used, actual cost may be direct costs plus overhead or may be direct cost plus overhead plus profit.

Allowable Cost

The owner's absolute maximum project cost, based on the project business case, which is the subject of the validation study. The allowable cost includes all elements: direct costs, overhead, and profit (also called ICL).

Big Room

A space where all stakeholders in the team can come together and work, typically with visual documentation posted. Shared space can support communication and dialogue, resulting in greater efficiency and work product that is updated in real time, as well as less reworking and revising. Big Room setup, duration, and usage varies.

Building Information Model(ing) (BIM)

The product (model) and process (modeling) of generating and managing building data during the life cycle of a building. BIM uses three-dimensional building modeling software. BIM includes building geometry, spatial relationships, geographic information, and quantities and properties of building components.

Blended Rate

An average hourly rate that can be used for financial tracking when precise amounts are not needed. Typically used for trade partners when a range of hourly rates based on person-hours can be averaged to project costs. (See Appendix 23 for an example of how a blended rate was used to calculate costs based on hours.) Can also be used in situations when design partners may not wish to highlight differences in salaries for personnel who share the same job title. Averaging multiple people at the same title creates one rate that can be openly shared without revealing sensitive information.

Burn Rate

The rate at which project funds are expended. Typically tracked in a spreadsheet with budgeted versus actual cost for labor and materials, focusing on rate of expenditure over time. (*See Appendix 24 for an example of how the burn rate can be tracked.*)

Choosing by Advantages (CBA)

A structured decision-making system that compares the advantages of alternatives based on objective facts and transparently evaluated subjective preferences.

Co-location

Physically locating personnel in a single area, often referred to as the Big Room, to enable constant communication and integrated thinking, build relationships, and increase productivity. Co-location may be face-to-face 100% of the time or part-time. Virtual co-location, the commitment of the team to collaborate at specific dates and times through use of web-based collaboration technology, is another method of co-location.

Conditions of Satisfaction (CoS)

An explicit description by an owner and/or other members of the IPD team, stating all requirements that must be satisfied to deem the outcomes as successful. Distinct from a project charter, which typically focuses on team-behavioral goals. (*See Appendix 2 and Appendix 4 for examples of CoS.*)

Dashboard

Visual management system to track data and metrics important to the team, which highlights whether the project is on track and also prompts actions. (*See Appendix 14 for examples of dashboards and their use in the Big Room.*)

Design Assist

Builders providing design assist services offer suggestions, insight, costing, and constructability review, but do not take responsibility for design, which remains with a design professional unaffiliated with the builder. All builders that are within the IPD group typically provide design assist or design/build services. In some instances, trade contractors who are not in the IPD group may provide design assist services under their subcontracts.

Design/Build

Design/build can refer to a project delivery method or a method for delivering an element of a project, such as a mechanical system. As a project delivery method, the design/builder is responsible for the design and construction of the project. As a method for delivering an element of a project, the design/builder has the design and the construction responsibility for that element. Fire protection systems, for example, are often delivered as a design/build element within an IPD project.

Design Management

Design management brings order and structure to the development of the design through defining outcomes and decision-making processes and by identifying and optimizing information flow and pull planning.

Guaranteed Maximum Price (GMP)

A cost-type contract that compensates the contractor for actual costs incurred plus a fee subject to a ceiling price.

Huddle

Huddle (or "daily huddle") is a very short daily stand-up meeting that addresses the day's work. Huddles are a part of scrum but are also frequently used in lean construction. (*See also Scrum.*)

Incentive Compensation Layer (ICL)

The team's collective, at-risk profit. The ICL can increase or decrease based on the project outcome. An adjusted ICL is the ICL after adjustment based on project outcome.

IPD Agreement or Integrated Form of Agreement (IFoA)

In this guide, we use IPD agreement to reference the multi-party or poly-party agreement that includes, at minimum, the owner, design professional, and constructor as signatories to the same construction contract. Examples include custom agreements (such as those by the law firm Hanson Bridgett) and templates (such as CCDC-30, ConsensusDocs 300, and AIA-C191 or C195). An IPD agreement is synonymous with IFoA. An IFoA or IPD agreement may be a multiparty (three-party agreement) or a poly-party agreement that can have more than three parties. (*See also Multiparty Agreement and Poly-party Agreement.*)

Integrated Project Delivery (IPD)

IPD is a contractually based approach, which creates an environment that enhances collaboration, innovation, and value. IPD is characterized by early involvement of IPD team members, shared risk and reward based on project outcome, joint project management, liability reduction among IPD team members, and joint validation of project goals.

IPD Team

The IPD team is made up of the participants who have placed their profit at risk and have the opportunity for increased profitability, based on project outcome. Under a multiparty agreement, IPD team members who are not signatory to the multiparty agreement are engaged through appropriate subcontracts or subconsulting agreements that reflect the terms of the multiparty IPD agreement. Sometimes called the risk/reward team, parties, or the ICP participants.

Last Planner System (LPS)

The collaborative, commitment-based planning system that integrates pull planning, make-ready look-ahead planning with constraint analysis, weekly work planning based on reliable promises, and learning based upon analysis of PPC and reasons for variance. (*See Appendix 15 for an example of LPS statistics.*)

Lean

A culture based on a set of principles focused on creating more value for the customer through elimination of waste, streamlined processes, and continuous improvement (*See More Resources for more information on lean.*)

Level of Development (LOD)

The LOD specification is a product of the BIMForum. Based on the basic LOD definitions developed by AIA, it is used to clearly define and communicate to what level of completion work will be done in a BIM and by whom: who will be responsible for modeling which building elements to a specific level of detail at a particular point in time. (*See Appendix 13 for an example of a LOD matrix.*)

Likert Scale

A common means of psychological measurement used to gauge a person's opinions, values, and/or attitude along a range of responses. The range of responses usually consists of five to seven possible answers—for example, ranging from strongly disagree to strongly agree—with a number value corresponding to each response.

Logs/Registers

This family of tools includes constraint logs and risk and opportunity registers. These have multiple functions. They are used to track and mitigate risks and issues. The development and consistent usage of them builds team consensus and can drive accountability. (*For examples of logs and registers, please see Appendix 10 and Appendix 11.*)

MEP

Mechanical, electrical, and plumbing systems. These are often inclusive of fire protection and data cabling as well.

Milestone

An item on a master schedule that defines the end or beginning of a phase or a contractually required event.

Multiparty Agreement

Referencing a three-party IPD agreement between owner, designer, and builder. Though the prefix multi does not imply a specific number, it is industry standard that multi-party is a three-party agreement due to the history of the development of IPD agreements. (*See also IPD Agreement and Poly-party Agreement.*)

Non-Signatory

A company that is participating in the project that is not part of the IPD team. That is, they are not included in the IPD agreement with the shared risk/reward and other terms.

Off-Boarding

The deliberately planned process for removing team members or firms.

On-Boarding

The deliberately planned process for bringing new players onto the team. In IPD, there is a need to on-board and align the initial team and to have a process for on-boarding new players added later to the team.

One-Piece-Flow

A methodology used to address a process from end to end with all parties involved in order to identify which step(s) must be completed for the next step to occur without waiting or waste.

Overhead (Home Office Overhead)

The amount, which may be expressed as a percentage applied to costs or a fixed amount, to compensate a firm for items such as rent, executive salaries, and other non-project-specific costs. (*To see an example of how overhead can be calculated, see Appendix 27 for trade partners and Appendix 28 for designers.*)

Owner Controlled Insurance Program (OCIP)

An OCIP is an insurance program in which the owner obtains a policy to cover loss and liability during the project, reducing the coverages provided by other parties, such as the construction manager/general contractor and trade partners. An OCIP program has requirements for safety management, reporting, and the like, which must be incorporated into the IPD team's plan.

Owner's Project Requirements (OPR)

Developed by the owner, this is a project narrative defining the owner's requirements. The OPR is often used as a basis for the team to develop the CoS. In the context of a high-performance certification, this can include quantitative measures, such as meeting LEED or Petal standards. (*See Appendix 3 for an example of OPR.*)

Percent Plan Complete (PPC)

A basic measure of how well the planning system is working, calculated as the number of commitments completed by the time stated divided by the total number of commitments made for the time stated. It measures the percentage of assignments that are 100% completed as planned. (*For examples of how PPC is visually tracked, see Appendix 14 and Appendix 15.*)

Plan-Do-Check-Act (PDCA; also sometimes Plan-Do-Check-Adjust)

A four-step process intended to support continuous improvement in a product or process: plan, do, check, act. This is conceived of as a repeating and never-ending cycle, which creates a feedback loop for teams to assess their ability to achieve and improve outcomes.

Plus/Delta

Performed at the end of an activity, such as a meeting or a decision process. This review is used to evaluate the activity. Two questions are asked and discussed. Plus: what produced value during the session? Delta: what could we change to improve the process or outcome?

Poly-party Agreement

An IFoA that has more than three parties and generally includes, as parties, all members of the IPD team. The distinction between a multiparty (three party) and poly-party agreement is relevant to contract structure, governance, and insurance.

Project Charter

(*See also Conditions of Satisfaction.*)

Project Implementation Team (PIT)

PITs are nimble, multidisciplinary groups of project participants assigned by the PMT to conduct deep dives into specific project needs (e.g., building envelope, mechanical systems). PITs typically have an initial mission, a time frame in order to perform their work and report back, and the authority to incorporate the right people to perform the work. These are sometimes called clusters or cluster groups. PITs can include all members of the team— PMT, signatories, non-signatories, owners, architects, contractor, trades, and suppliers. Common PITs include structure, mechanical, electrical, envelope, etc. The specific number of PITs needed will be determined by the team. (*See also Project Management Team.*)

Project Management Team (PMT)

A team composed of representatives from each IPD contract party, with membership as defined by the specific IPD contract and subsequently others as jointly agreed by the parties. The PMT is charged to act in a collaborative manner to provide project management leadership during the design and construction process in a concerted effort to achieve the project's objectives. The PMT is the project's administrative workhorse, making the tough decisions and monitoring financials. Sometimes called the core group or core team. Interfaces with the SMT and PIT. (*See also Senior Management Team and Project Implementation Team.*)

Project Team

The totality of all firms participating in the project, regardless of their status in the risk/reward structure. For the purposes of this guide, the firms participating in risk/reward make up the IPD team. There may be firms working on the project that are not part of the risk/reward structure. These are referenced as non-signatory or the project team. The totality of all the individuals on the team is referenced as project participants. (*See also IPD Team and Non-Signatory Agreement*).

Pull

A method of advancing work when the next-in-line partner is ready to use it. A request from the partner signals that the work is needed and is pulled from the performer. In the pull method, work is released when the other members of the team are ready to use it.

Push

The opposite of pull. During push, an order is made from a central authority based on a schedule and advancing work based on a central schedule. Releasing materials, information, or directives possibly according to a plan but independent from whether or not the downstream process is ready to process them.

Request for Information (RFI)

A formal question asked by one party of the contract to another party. Typically, a request from the contractor to the designer.

Request for Proposals (RFP)

Owner's call for teams to submit proposals. In IPD this often includes how the team is going to handle collaboration and integration. (*See Appendix 1 for an RFP example.*)

Request for Qualifications (RFQ)

Typically includes relevant previous work, key personnel, and approach to work. In IPD this often includes demonstrations of lean and IPD experience.

Risk/Reward

A collectively agreed upon amount or percentage of final cost that will be distributed among the members of the IPD team (sometimes called risk/reward pool) if project goals are met. Sometimes called ICL or profit pool.

Rough Order of Magnitude (ROM)

Estimate of time or cost before details are known. A way to describe the impact and likelihood of an occurrence that could impact the project budget, positively or negatively. Calculated by taking possible cost or savings multiplied by the probability of occurrence. Typically used with risk logs or opportunity logs, sometimes combined into one format, sometimes weighted with probabilities and costs so that it can be managed in conjunction with contingency funds.

Scrum

Scrum is a term borrowed from agile project management, often used in software development, referring to a process involving small teams engaging in short, repeatable, sustainable "sprints," the outcome of which is a chunk of delivered value.

Senior Management Team (SMT)

A team composed of representatives from each IPD team member, typically the project executive of the firm. The SMT always handles dispute resolution and backs up the PMT as required. In many cases they also conduct contract negotiations and resolve questions of scope change, but this can alternatively be done by the PMT. The SMT is composed of one C-level executive from every party who signs the IPD agreement.

Target Cost (TC)

The cost goal established by the project team as the target for its design and delivery efforts, typically determined after the validation process. In some projects, there is only TC, which can be adjusted by the owner in the rare situations when that is appropriate. Other times, TC is broken into two measures:

> Base Target Cost: The TC amount that matches the base program in the project objective.

> Final Target Cost: The TC amount that matches the base program, plus any value added Items. Because the value added Items are funded from savings off of the base TC, the final TC must be less than or equal to the base TC (unless there are change orders).

Target Value Design (TVD)

A disciplined approach to design that requires project values, cost, schedule, and constructability to be basic components of the design criteria, and uses cost targets to drive innovation in designing a project to provide optimum value to an owner. TVD uses constructability and cost information from the owner and IPD team before design decisions are made to allow the design to progress within the base TC, final TC, and schedule. (*To see an example of PIT tracking during TVD, see Appendix 26.*)

Trade Partners

Trade partners are the IPD team members (signatories to the IFoA) who are the specialty contractors engaged to put the construction work in place. Trade partners typically include mechanical contractor, electrical contractor, structural steel contractor, and the like. Not to be confused with trade contractors, subcontractors, and trades, which are ambiguous terms that do not imply membership on the IPD team.

Validation

Validation is a process through which the IPD team establishes certainty for the project. It proves or disproves whether the team can meet the full range of the owner's CoS within the owner's constraints (including cost and schedule). Validation is not compressed schematic design. The project is developed only to the degree necessary to achieve certainty. Validation is a go/no-go gate, under-taken at the beginning of the project, and often has its own budget, schedule, prerequisites, and approvals. (*For an example of a validation checklist, see Appendix 8.*)

Virtual Design and Construction (VDC)

The use of BIM and other tools to optimize and coordinate design, virtually rehearse and manage construction, and/or operations.

Visual Management

Placing tools, parts, plans, schedules, measures, and performance indicators in plain view for transparency, allowing the system to be understood at a glance by everyone involved and actions taken locally in support of system objectives. (*For examples of dashboards and visual management in the Big Room, see Appendix 14.*)

Weekly Work Plan (WWP)

The commitment-level step of LPS, identifying the promised task completions agreed on by the project team. The WWP is used to determine the success of the planning effort and to determine what factors limit performance and is the basis of measuring PPC. (*See Appendix 16 for examples of WWPs.*)

APPENDICES

The documents selected for inclusion in this section are a small sample of the myriad of possible ways to tackle project issues. A few examples were selected for their unique language or format (e.g., 01_Request for Proposals, 02_Owner Goals and Conditions of Satisfaction, and 07_Organization Chart). The majority were chosen because they contain typical content and/or use a typical format found on IPD projects.

The samples in this guide are intended to stimulate thought and give project leaders a sense of the scope and expanse of tools that teams have found helpful on IPD projects. IPD teams need to be creative and inventive. Teams should ask: what is the problem we need to solve, and what specifically do we need to solve it?

Documents were generously provided by a wide range of people with extensive IPD experience. In most cases, identifiable information was redacted to preserve confidentiality.

RE: INTEGRATED PROJECT DELIVERY OF Project Name

The   family of companies   includes   Company X   one of the largest and most trusted   design consulting firms in North America. X's vision is to provide a better way by design and they strive to awesomize the experience of their business partners by honouring the core values of their corporate culture. Those values include: trust, balance, purpose, teamwork and the pursuit of perfection. This translates into an exceptional customer experience, unlike any other in the industry.

CLICK HERE FOR THE X STORY

Contractor A is the leading mid-sized contractor in Canada focused on providing an exceptional customer experience. We believe that there's a better way to build and as a result we see things differently. Tailored. Unique. Smart. Ours is a process that breeds inspiration and confidence. It takes a distinct perspective to see things in a new way. That's why companies engage Contractor A to not only bring their vision to reality but to help shape it though an intelligent and collaborative approach to construction.

CLICK HERE FOR Contractor A's STORY

Early this year, X engaged   A   as prime contractor for a new 30,000 square foot office building in Edmonton. The project is deep green. It is intended to serve as a living case study of a better way to develop, design and build. The current team believes that integrated project delivery is that better way. There are many who talk about this. There's lots of smoke but no fire. Our team wants to light that fire on this project. In order to do that, we need your help. That's why you've been invited to participate.

CLICK HERE FOR THE IPD Project A STORY
CLICK HERE FOR THE IPD Project B STORY
CLICK HERE FOR THE IPD Project C STORY

We're only interested in working with partners who believe in this. If that's you, we want you to come-in and meet with us. During that meeting, we'll show you how gain/pain sharing could work. Then, we'll ask you to give us a proposal that tells us how you think you can add value to the project. We'll shortlist and the team will decide if your firm is a good fit. We'll need a commitment from you to go open book with us so we can help reduce your costs. To start the conversation, we need to know the following:

WHAT __% PROFIT WOULD YOU LIKE TO MAKE?
WHAT __% OVERHEAD DO YOU NEED TO RECOVER?
WHAT'S YOUR __% BURDEN ON LABOUR?

Contractor B   is based in San Francisco and has done more IPD than all others in North America. They have agreed to provide training to our project team. That training will likely involve a trip to San Francisco so that we can observe design meetings occurring real-time in an integrated manner. Then, the team from B will deliver a one day workshop intended to set us off on the right path. The cost related to this training will be shared by the member companies of the integrated team.

So what's in it for you? Firstly, this method of project delivery is the future of our business. This is an opportunity for you to lead your industry. Secondly, this project is a living case study. We will be documenting our journey with various forms of media that can be used by the team members. Thirdly, this process provides lower risk and higher opportunity than traditional delivery methods. Complex projects are challenging. That doesn't need to be made worse by antagonistic behavior and us vs. them thinking.

01_Request for Proposals

This request for proposal (RFP) was issued for the first IPD project in this market. The concise one-page document covers context of the project, background about IPD as a delivery method, and outlines what information is needed. Overall, it captures the enthusiasm and energy around IPD as a better way to build, and it provides a welcoming invitation to join a journey supported by trusted experts.

CONDITIONS
OF SATISFACTION

(developed by design team in IPD worksession on February 20, 2013, updated April 2, 2013)

PHASE I — CONCEPTUAL DESIGN

- An aesthetically beautiful building design that inspires donors
- Performance metrics that inspire the project team and donors
- Flexible fund-raising opportunities developed collaboratively by the team
- Team integration and alignment on performance metrics
- The right team members attend the right meetings at the right time
- A high-performing lean team that fulfills commitments on time
- A team that has fun, and throughout the project, remains friendly and respectful

PHASE II — FINAL DESIGN / CONSTRUCTION

- A beautiful, high-performance building that is embraced by and enlivens Basalt & RMI
- A building that is replicable in the marketplace
- Town-of-Basalt approval of a building design that works for RMI
- RMI site unencumbered by 100-year floodplain
- Fair profit for all team members
- Formal project-team agreement to "integrated project delivery" at Phase 2 launch
- Zero "change orders" (CO's) and no unnecessary "requests for information" (RFI's)

02_ Owner Goals and Conditions of Satisfaction

This page is an excerpt from a larger document developed by the IPD team to summarize their goals. It includes project-specific goals (both qualitative and quantitative) as well as team goals for behavior and effectiveness. An unusual goal was to make the project a model for other relatively small-sized buildings to use IPD in order to meet extraordinarily high-performance goals. (*See Appendix 3 for how this condition of satisfaction relates to an owner's program of requirements.*)

1 Executive Summary

1.1 Project Objectives

The intent of this document is to systematically and clearly communicate the expectations of the owner to the design and construction team. The objective of the OPR is to provide the base information required by the design and construction teams to deliver a well-planned, cost-effective building that meets the following goals:

1. Create a building that exemplifies RMI's mission and program – through its story it educates visitors, enhances RMI's convening power, heightens RMI's visibility and creates space for collaboration and learning.
2. Create the highest performing building possible – a building that moves beyond efficiency and net zero to a restorative building, contributing positively in all aspects of energy, water, transportation, materials and resource use.
3. Create a replicable process and business case – integrated design process that translates across all project types while driving competitive operational costs.
4. Create a beautiful structure focused on community outreach and occupant experience – project should inspire and express the RMI mission through craft and aesthetics, durability, functionality, support of collaboration, research and programmatic flexibility.

1.2 Developing the OPR

It is the owner's responsibility to develop an accurate OPR to provide the designers and constructors with the correct information to deliver a building that meets the owner's desires. The OPR forms the basis from which all design, construction, acceptance and operational performance evaluations are made. This is a living document that will be updated throughout the design and construction process.

1.3 BOD Responses

The Basis of Design is a document created by the designers that records the concepts, calculations, decisions and product selections used in the design to meet the goals of the OPR and to satisfy the applicable regulatory requirements, standards, and guidelines. The BOD document includes both narrative descriptions and specific assumptions made by the designers. The early design vision book is serving as the project's BOD

1.4 Project Team Information

Provided in Appendix A.

03_Owner's Project Requirements

Executive summary from a fifty-five-page document that preceded the development of the conditions of satisfaction. (*See Appendix 2.*) In this case, the owner, Rocky Mountain Institute, was highly expert in advising building owners about ways to translate their business objectives into project requirements and set net zero as a goal. This Owner's Project Requirements contained a project description that included the size, location, budget, schedule, contingency, future expansion, codes and standards, building operation, owner's risk tolerance, warranty requirements, and delivery method. Another section outlined the owner's project requirements for sustainability goals; mechanical-systems comfort criteria; commissioning requirements; accessibility; envelope; building dashboard; mechanical, electrical, and plumbing requirements; acoustic requirements; modeling process; energy and thermal modeling software; life-cycle cost analysis; measurement and verification; and training and handover documents.

Keep an Open Mind

We will consider all options and methods to achieve our goals. We commit to continuously improve the process by promoting a free and safe environment where new ideas can be shared at all times.

Consider all options and methods to achieve our goals

Being open to ideas

Continuous improvement

Understanding and Learning

Building on past projects to help team members achieve leading edge skills to grow in their understanding of IPD success.

Develop new tools / processes for next building construction

Reflection from past projects

Learning, understand working effectively in IPD

Growth and understanding of IPD

Fun

We want to have fun engaging together as a team, building relationships, and enjoying the process of delivering a successful project.

Team engagement

Project team has fun while delivering project

A project team that has fun and builds relationships

Positive team morale

Conditions of Satisfaction *(7)*

Efficient Design

We work efficiently to develop an accessible building that has an effective design for the end users (use and operations) using a constructable approach.

Cost effective and efficient use of time and materials

Constructability

Efficient and effective design (flow and operations)

Ease of operations (not overly complex)

Flexibility, Durability, and Life Cycle

We are forward thinking. We consider a successful building one that plans for optimal performance and future use, while minimizing the life-cycle costs of our project.

Durability in material and equipment choice

Life cycle of the building is always considered

Adaptability in design

Building meets present requirements and is flexible enough for the future

04_Project Values

The IPD team convened a project-values workshop early in the process, and this list documents the outcome. The particular list from which this excerpt is from identifies and describes the following: five behavioral values (ranging from communication to fun), seven conditions of satisfaction (ranging from how the team will be effective to the project outcomes), and three key performance indicators (schedule, cost, safety). For each item on the list, measures of success, descriptors of what success looks like, and guiding principles are named.

Partner		People/Team			Engagement with this Opportunity			Commercial and Community		Totals
		Qualifications of individuals relative to project scope, IPD and Lean	Perception of collaboration, team interaction and chemistry	Willingness to learn	Innovation	Target Value Design and costing approach	Feedback regarding schedule	Commercial terms	Diversity program (MBE, WBE, local)	Totals
	Weight	20%	15%	10%	10%	20%	10%	10%	5%	
Name:	Score (0-10)	7	8	8	7	7	6	4	9	
Contractor A	Weighted Value	1.40	1.20	0.80	0.70	1.40	0.60	0.40	0.45	6.95
Name:	Score (0-10)	9	8	8	8	8	5	7	4	
Contractor B	Weighted Value	1.80	1.20	0.80	0.80	1.60	0.50	0.70	0.20	7.60
Name:	Score (0-10)	6	6	7	7	7	5	5	2	
Contractor C	Weighted Value	1.20	0.90	0.70	0.70	1.40	0.50	0.50	0.10	6.00

05_Partner Evaluation Matrix

Selecting the IPD team can be supported by tools such as this example. Selection criteria can be publicized in the request for proposals, listed on a matrix, and assigned a weight. Those team members evaluating the candidates can record their scores and tabulate the results. In general, selection processes that use scoring have been shown to reduce implicit bias. Assigning values and weights to qualitative attributes, such as willingness to learn, can be tricky—evaluators should agree on what evidence should be used to assign a score. Criteria is organized into three broad categories: people/team, project-specific engagement, and commercial. Contractor B's score, circled in green, is the highest.

Task Description	IPD Management Team		IPD Agreement Parties						Non-ICL Participating Consultants	Non-ICL Participating Subcontractors	
	SMT	PMT									
3.0 IMPLEMENTATION DOCUMENTS PHASE											
3.1 Project Administration Services											
.1 Provide overall facilitation, coordination, organization, and direction of PMT	S										
.6 Verify/Update standards for BIM and digital coordination		P	C	C	C	C	C	C			
.7 Execute BIM Plan			C	P	C	C	C	C		DA	DA
.8 Provide required legal and insurance		S									
.9 Measure Project Goal compliance through tracking of metrics		P	C	C	C	C	C	C			
.10 Confirm that all necessary Work is accounted for		S									
.11 Assign responsibility for other Project Administration services		S									
3.2 Data Gathering/Programming/Regulatory Agency Services											
.1 Agency Consultation/Review/Approval			C	P	C	C	C	C	C		
.2 Facilitate final user reviews and approvals			P	C	C	C	C	C			
.3 Initiate transition planning to utilize completed Project			P	C	C	C	C	C			
.4 Coordinate complete information for legal requirements of Project as it relates to Owner's Procurement Method		S									
.5 Complete information required for procurement, assembly, layout, detailed schedule, and procedural information (testing, commissioning)			C	P	C	C	C	C		DA	DA
.6 Manage and lead strategy regarding negotiations with jurisdiction providing permits		P	C	C	C	C	C	C			
3.3 Cost and Schedule Validation Services											
.1 Coordinate financial reqs. that are necessary to begin construction		S									
.2 Provide continuous cost feedback using available information; all item quantities to be based on quantities exported from the Model, quantity surveys, or lump sums provided by subcontractors and suppliers			C	C	C	P	C	C		DA	DA
.3 Verify schedule for long lead items			C	C	C	P	C	C		DA	DA
.4 Finalize construction schedule			C	C	C	P	C	C		DA	DA
.5 Provide schedule for application submittals and review completion			P	C	C	C	C				
.6 Finalize construction cost			C	C	C	P	C	C		DA	DA
.9 Ensure finalization/end user approval of detailed phasing plan that supports Operational and Code Requirements			C	C	C	P	C	C			
.7 Assign responsibility for other cost and scheduling services		S									
4.0 CONSTRUCTION PHASE											
4.1 Project Administration Services											
.1 Provide overall facilitation, coordination, organization, and direction of PMT	S										
.2 Provide overall facilitation, coordination, and direction of PIT		S									
.3 Confirm PIT compliance with Project Requirements		S									

LEGEND
S - Sole Responsibility for indicated task
P - Primary Responsibility the indicated task. Parties with Primary responsibility are responsible for task but assisted by others
C - Contributing Responsibility
DA - Design Assist Responsibility
DB - Design Build Responsibility

06_Responsibility Matrix for an IPD Agreement

While not every IPD agreement uses a responsibility matrix, it can be one of the key exhibits jointly developed by the IPD team during validation. The responsibility matrix allocates tasks during the different phases of the project among the team and clarifies roles, responsibilities, and the transfer of responsibility as the project develops. This responsibility matrix allocates tasks to both members of the IPD team (blue) and non-signatory consultants and subcontractors (pink). This example is excerpted from a 238-line spreadsheet.

SCHOOL OF ENGINEERING

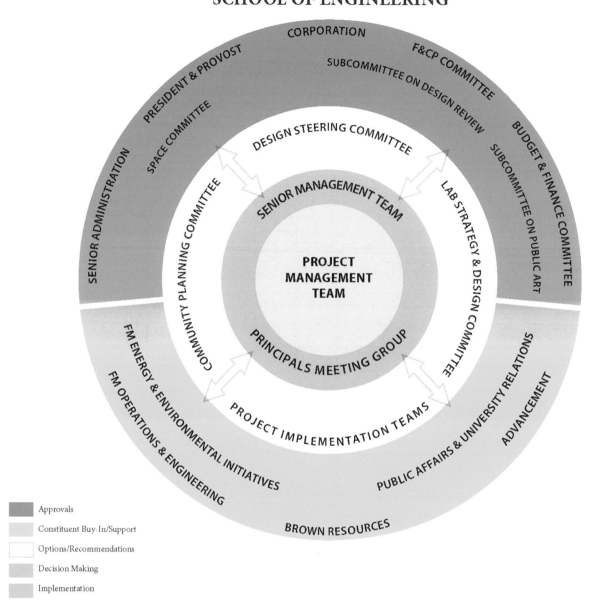

Legend:
- Approvals
- Constituent Buy-In/Support
- Options/Recommendations
- Decision Making
- Implementation

07_Organization Chart

A large public owner with complex approval and communication layers created this unusual take on a standard organizational chart. This chart describes the primary roles and responsibilities of the project management team, senior management team, project implementation teams, and other groups, committees, and subcommittees that influence the project, while also identifying who is responsible for project approvals, decision-making, and implementation. This graphic is an excerpt from a single ledger-sized sheet. In the full document, individual members for nine of the primary stakeholder groups are listed, and the responsibilities of eighteen stakeholder groups are described relative to campus design, construction, and operations.

Courtesy of Michael Gugliemo.

08_Making a Table of Contents for Validation Example A

This TOC for a Seniors Care project shows a relatively typical table of contents for a validation report. Decisions on what to include and what not to include in the validation report is an important step in establishing the scope of the validation process.

Item	Description	< $10M			$10M to $29.99M			$30M to $100M			> $100M		
(check applicable project cost)		☐			☐			☐			☐		
		Mandatory	Included	N/A	Mandatory	Included	N/A	Mandatory	Included	N/A	Mandatory	Included	N/A
1. Introduction													
1.1	Cover Sheet		□	□		□	□	⊠			⊠		
1.2	Table of Contents		□	□	⊠			⊠			⊠		
1.4	Executive Summary		□	□		□	□	⊠			⊠		
2. Program, Planning and Design													
2.1.b	Project Charter (A3.1)	⊠			⊠			⊠			⊠		
2.1.c	Model of Care		□	□		□	□		□	□	⊠		
2.2.a	Operational Assumptions	⊠			⊠			⊠			⊠		
2.2.b	Service Plan Summary		□	□		□	□		□	□		□	□
2.2.c	Regulatory constraints/requirements	⊠			⊠			⊠			⊠		
2.3.a	Stacking Diagrams	⊠			⊠			⊠			⊠		
2.3.b	Department Block Layouts	⊠			⊠			⊠			⊠		
2.3.d	Concept floor plan(s) & alternatives	⊠			⊠			⊠			⊠		
2.3.g	Detailed room design/room data sheets for key rooms	⊠			⊠			⊠			⊠		
2.3.h	Design options/set-based studies		□	□	⊠			⊠			⊠		
2.3.i	Simulation Results		□	□		□	□		□	□		□	□
2.3.j	Code Research	⊠			⊠			⊠			⊠		
2.4.a	Site/Urban Plan Analysis/Context		□	□		□	□		□	□	⊠		
2.4.b	Utilities & Encroachments		□	□	⊠			⊠			⊠		
2.4.c	General Site considerations		□	□	⊠			⊠			⊠		
2.4.d	Site Survey/key features/impacts	⊠			⊠			⊠			⊠		
2.4.e	Traffic/Access/Parking		□	□	⊠			⊠			⊠		
2.4.g	Concept Site Plan	⊠			⊠			⊠			⊠		
2.4.h	Building Massing Options		□	□	⊠			⊠			⊠		
2.4.i	Building Concept Elevations		□	□	⊠			⊠			⊠		
2.4.j	Exterior wall studies/sections		□	□		□	□	⊠			⊠		
2.4.k	Concept Roof Plan		□	□	⊠			⊠			⊠		
2.4.p	Outline Specifications		□	□	⊠			⊠			⊠		
2.4.q	Sustainability strategy/tracking	⊠			⊠			⊠			⊠		
3. Building Systems													
3.1	Structural Systems		□	□		□	□	⊠			⊠		
3.2	Exterior Skin Systems		□	□	⊠			⊠			⊠		
3.3	Mechanical Systems	⊠			⊠			⊠			⊠		

08_Making a Table of Contents for Validation Example B

This is an excerpt from a standard form used by an owner with the same project type of multiple scales. Using this checklist, the IPD team can generate a table of contents for a validation report, which lists the most important steps in establishing the scope of validation. This particular checklist emphasizes different factors depending on the size and scope of the project. The excerpt hides some line items; sections not included in this view are: building systems, risk assessment, schedule and budget, and appendices.

DECISION MATRIX

PIT: ████████████

Project values will be used to guide the team in decision making. Use this matrix on any major decision document that grades the decision on its affect (red, yellow, green) on the overall project values. Where there is a conflict between values, the document should discuss how the conflict will be resolved. If a decision doesn't affect a value, the team should question the necessity of the action.

Decision Outline

NOTES

	EFFECT of DECISION			
	POS	NEU	NEG	N/A
SUPPORTS EDUCATIONAL MODEL				
INSPIRATIONAL				
SUSTAINABLE				
OPERATIONAL				
BUILDABLE				
LEGACY				

DECISION MADE   (+ ANY BACKUP)

Budget estimate by Team _____

Budget impact validated by GC _____

DIR #
Please indicate in cell B33 if there is a Deisgn Information Report associated with this item

COMPLETED BY: _____

DATE: _____

EMAIL to:

Please indicate in cell F37 if Decision was to "Accept", "Reject" or "Under Review"

09_Decision Matrix

A decision matrix holds team members accountable for the values agreed to at the outset. Cost is one criterion, but overall project values guide decision-making. This decision matrix evaluates a decision based on its effect (e.g., positive, neutral, negative) on project values. Often, the effect is positive for one value but negative for another; discussion to resolve helps to clarify priorities and align goals. If a decision did not affect value, the necessity of the action may be questioned. This tool is sometimes employed instead of an A3.

RISK / OPPORTUNITY LOG

UPDATE DATE: 4/9/18

DG	OPPORTUNITY THAT WILL IMPROVE CONDITIONS
LG	OPPORTUNITY THAT MIGHT IMPROVE CONDITIONS
O	MEDIUM RISK
Y	HIGH RISK
R	CRITICAL - HIGHEST RISK

	RISK / OPPORTUNITY IDENTIFICATION				RISK / OPPORTUNITY ASSESSMENT		MANAGEMENT PLAN		
ID	Certainty / Uncertainty	Topic	Description	Category	Probability: Low/Med/Hi	Impact: Low/Med/Hi	Action	By Whom	By When
1	DG	Big Room	Co-locating for expedited documentation and approval	Schedule	High	High	Determine location and start date for Big Room	CM - ■	4/30/18
2	R	Onsite labor	Local electrical labor shortage	Schedule	High	Medium	Prefab offsite as much as possible: Room in a box; headwalls; bathrooms?	Arch - ■	4/21/18
3	LG	Phased permitting	Early site, concrete, superstructure	Cost	Medium	High	Contact AHJ to discuss permitting options	CM - ■	4/9/18
4	Y	Design team staffing not complete	Room strategy and responsibility matrix without full staffing plan	Team	Medium	High	Plan by end of month	Arch - ■	4/27/18
5	Y	Final approval of layouts	Owner team final approval ongoing	Schedule	Medium	Medium	Approval by end of week	Owner - ■	4/13/18

10_Risk/Opportunity Log

The risk and opportunity log is organically developed by the team based on what the team is willing and able to manage. This log can be extensive and highly detailed, or short and simple, depending on what the team decides. While the general purpose of this document is to track risks and opportunities, the process of the creating the document aligns the team around their capacity and willingness to manage the project.

The log example here uses red, yellow, and green to color code the uncertainty level of risks and opportunities. The log identifies the risk and its level of uncertainty, assesses its probability and impact on savings, and puts into place a risk-management plan, with a team member ready to take action. This sample is fairly simple and focused on the essential elements. Some teams have a more extensive spreadsheet with fine grained information.

Constraint Log

At the OAC meeting, we're making sure ■ has it on his radar to approve the rebar submittal in enough time to get the material on site for our currently scheduling installation date.

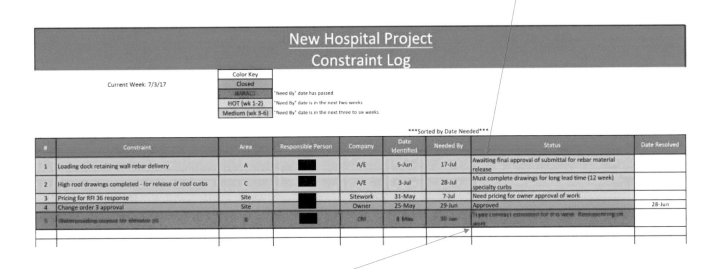

New Hospital Project
Constraint Log

Color Key	
Closed	
IMPACT	"Need By" date has passed
HOT (wk 1-2)	"Need By" date is in the next two weeks
Medium (wk 3-6)	"Need By" date is in the next three to six weeks

Current Week: 7/3/17

Sorted by Date Needed

#	Constraint	Area	Responsible Person	Company	Date Identified	Needed By	Status	Date Resolved
1	Loading dock retaining wall rebar delivery	A	■	A/E	5-Jun	17-Jul	Awaiting final approval of submittal for rebar material release	
2	High roof drawings completed - for release of roof curbs	C	■	A/E	3-Jul	28-Jul	Must complete drawings for long lead time (12 week) specialty curbs	
3	Pricing for RFI 36 response	Site	■	Sitework	31-May	7-Jul	Need pricing for owner approval of work	
4	Change order 3 approval	Site	■	Owner	25-May	29-Jun	Approved	28-Jun
5	Waterproofing at west site elevation pit	B	■	CM	8-May	30-Jun	Type contract extended for this week. Resequencing on work	

Due to initial bids coming back too high, the CM is forced to extend the bid period for the waterproofing contractor. The original schedule called for waterproofing to be installed this week. The team will either decide to add this specific waterproofing scope as a change to someone else's contract who is already on site, or resequence the waterproofing at this location to a later date.

11_Constraint Log Example A

This tool tracks issues over time, anticipating those that could delay the project. Identified issues are described, responsible individuals are named, and status updated. Some teams chose to integrate constraint logs into their weekly work plan; others use a separate document as shown here.

Behavioral Health Phase 2

CONSTRAINT LOG

ID	DATE ADDED	CONSTRAINTS (six weeks)	ACTIVITIES DELAYED	OSHPD ELEMENT	REQUESTOR	PERFORMER	DATE REQ'D	PROMISE DATE	COST IMPACT	STATUS	DAYS OPEN	CLOSE DATE
0014	7/24/14	5172.2 Shower Room: Surface Mounted 1x4 conflicting w/sprinkler coverage; ▮▮▮ to confirm if light fixture conflicts w/sprinkler	Install Light Fixture	no	▮▮▮	▮▮▮	07/31/14			CLOSED	7	7/31/14
0013	7/17/14	Signage at 5163, 5167A, & 5168A: signage needs to be corrected as required by ▮▮▮	Occupancy	yes	▮▮▮	▮▮▮	ASAP			CLOSED	8	7/25/14

ID	DATE ADDED	ISSUES (long term)	ACTIVITIES DELAYED	OSHPD ELEMENT	REQUESTOR	PERFORMER	DATE REQ'D	PROMISE DATE	COST IMPACT	STATUS	DAYS OPEN	CLOSE DATE
1003	7/1/14	New Location of Strobe in Gurney Shower Room:	Phase 2B FA		▮▮▮	▮▮▮	09/02/14			OPEN	38	
1002	5/5/14	Sprinkler Work Not Shown on Drawings: Room #5168, Stair 2, and ▮▮▮ Room all not shown to receive institutional heads on drawings but do need them. Need RFI/CO/ASI prior to inspection.	OSHPD Final	yes	▮▮▮	▮▮▮	06/18/14		yes	CLOSED	28	6/2/14
1001	3/28/14	OSP Numbers: need OSP numbers for panels	OSHPD Final	yes	▮▮▮	▮▮▮	06/01/14			CLOSED	76	6/12/14

11_Constraint Log Example B

Sample courtesy of Paulo Napolitano and Herrero Builders.

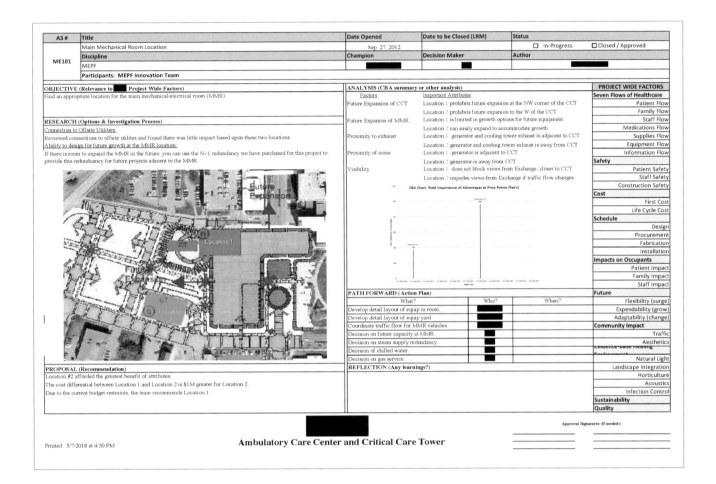

A3 #	Title		Date Opened	Date to be Closed (LRM)	Status	
ME101	Main Mechanical Room Location		Sep. 27, 2012		☐ In-Progress	☐ Closed / Approved
	Discipline		Champion	Decision Maker	Author	
	MEPF		███████	███	████████	
	Participants: MEPF Innovation Team					

OBJECTIVE (Relevance to ███ Project Wide Factors)
Find an appropriate location for the main mechanical/electrical room (MMR)

RESEARCH (Options & Investigation Process)
Connection to Offsite Utilities:
Reviewed connections to offsite utilites and found there was little impact based upon these two locations.
Ability to design for future growth at the MMR location:
If there is room to expand the MMR in the future, you can use the N+1 redundancy we have purchased for this project to provide this redundancy for future projects adacent to the MMR.

PROPOSAL (Recommendation)
Location #2 afforded the greatest benefit of attributes.
The cost differential between Location 1 and Location 2 is $1M greater for Location 2.
Due to the current budget restraints, the team recommends Location 1.

ANALYSIS (CBA summary or other analysis)

Factors	Important Attributes
Future Expansion of CCT	Location 1 prohibits future expanion at the NW corner of the CCT
	Location 2 prohibits future expanion to the W of the CCT
Future Expanion of MMR	Location 1 is limited in growth options for future equipment
	Location 2 can easily expand to accommodate growth
Proximity to exhaust	Location 1 generator and cooling tower exhaust is adjacent to CCT
	Location 2 generator and cooling tower exhaust is away from CCT
Proximity of noise	Location 1 generator is adjacent to CCT
	Location 2 generator is away from CCT
Visibility	Location 1 does not block views from Exchange, closer to CCT
	Location 2 impedes views from Exchange if traffic flow changes

CBA Chart: Total Importance of Advantages at Price Points (Fee's)

PATH FORWARD (Action Plan)

What?	Who?	When?
Develop detail layout of equip in room.	████	
Develop detail layout of equip yard.	████	
Coordinate traffic flow for MMR vehicles.	████	
Decision on future capacity at MMR.	██	
Decision on steam supply redundancy.	██	
Decision of chilled water.	██	
Decision on gas service.	██	

REFLECTION (Any learnings?)

PROJECT WIDE FACTORS

Seven Flows of Healthcare	
	Patient Flow
	Family Flow
	Staff Flow
	Medications Flow
	Supplies Flow
	Equipment Flow
	Information Flow
Safety	
	Patient Safety
	Staff Safety
	Construction Safety
Cost	
	First Cost
	Life Cycle Cost
Schedule	
	Design
	Procurement
	Fabrication
	Installation
Impacts on Occupants	
	Patient Impact
	Family Impact
	Staff Impact
Future	
	Flexibility (surge)
	Expendability (grow)
	Adaptability (change)
Community Impact	
	Traffic
	Aesthetics
Evidence-base Healing	
Environment	Natural Light
	Landscape Integration
	Horticulture
	Acoustics
	Infection Control
Sustainability	
Quality	

Approval Signatures (If needed):

Printed: 5/7/2018 at 4:50 PM

Ambulatory Care Center and Critical Care Tower

12_A3

A3s capture all the options that were studied by the team, recording final decisions and rationale. It can serve as a reference later if that rationale needs to be revisited. Developing the A3 can build consensus around the decision, since it gathers input and addresses concerns as it is circulated within the team. An A3 can also be used to simply document discussion of options, even if it turns out to be moot.

	MF Spec Section	Relevant	A100 Date LOD	ME	DD Date LOD	ME	CD Date LOD	MEA	As-Built Date LOD	MEA
Exterior Wall Specialties										
Exterior Horizontal			200	Arch.	200	Arch.	300	Arch.	300	Contractor
Roofing		B – Roof								
Steep Slope Roofing	7311 ASPHALT	B – Roof	200	Arch	200	Arch	300	Arch	300	Contractor
Roof Panels	7411 MANUFACTURED		200	Arch	200	Arch	300	Arch	300	Contractor
Low Slope Roofing	7319 POLYMERIC	B – Roof								Contractor
Protected Membrane Roofing	7530 EDPM SINGLE-		NM		200	Arch	300	Arch	300	Contractor
Protected Membrane Roofing	7531 TPO SINGLE-PLY		NM		NM		NM		NM	
Canopy Roofing			200	Arch	200	Arch	300	Arch	300	Contractor
Roofing Supplementary	7620 SHEET METAL		NM		NM		200	Arch	200	Contractor
Roof Appurtenances										
Roof Accessories (Ladders, curbs, vents, walkways, and snow...			NM		NM		200	Arch	200	Contractor
Roof Specialties (Cupolas, spires, steeples, and weathervanes.)			NM		NM		200	Arch	200	Contractor
Rainwater Management			NM		200	Arch	300	Arch	300	Contractor
Traffic Bearing Horizontal		B – Roof								
Traffic Bearing Coatings	Check with Jeff, Bob &								200	Contractor
Horizontal Waterproofing									200	Contractor
Wear Surfaces									200	Contractor
Horizontal Enclosure										
Horizontal Openings		B – Ext.								
Roof Windows and Skylights	7721 ROOF SKYLIGHTS		100	Arch	200	Arch	300	Arch	350	Contractor
Vents and Hatches	7720 ROOF SMOKE VENTS		NM		200	Arch	300	Arch	350	Contractor
Horizontal Openings			NM		NM		NM			
Overhead Exterior Enclosures		Ceilings (susp.								
Exterior Ceilings	7715 METAL CEILING		100	Arch	200	Arch	300	Arch	300	Contractor
Exterior Soffits			NM		200	Arch	300	Arch	300	Contractor
Exterior Bulkheads			NM		200	Arch	300	Arch	300	Contractor
INTERIORS										
Interior Construction			200	Arch.	200	Arch.	300	Arch.	300	Contractor
Partitions		C - Partitions							300	Contractor
Interior Fixed Partitions	9250 GYPSUM DRYWALL	A, B Cold Formed Metal Framing;	200	Arch	300	Arch	300	Arch	300	Contractor
Wood Framing	6100 ROUGH CARPENTE	C - Partitions								
Building Insulation	7200 BUILDING INSULA	C - Partitions	NM		100	Arch	100	Arch	100	Contractor
Interior Windows		C – Int. Windows								
Interior Operating Windows										
Interior Fixed Windows			200	Arch	300	Arch	300	Arch	300	Contractor
Fire Rated Windows	08801 FIRE RATED GLAS	C - Int. Windows	200	Arch	300	Arch	300	Arch	300	Contractor
Interior Special Function		C - Int. Windows								
Interior Window Supplementary			NM		200	Arch	300	Arch	300	Contractor
Interior Doors		C – Int. Doors								
Interior Swinging Doors	8110 STEEL DOORS		300	Arch	300	Arch	300	Arch	300	Contractor
Interior Entrance Doors			300	Arch	300	Arch	300	Arch	300	Contractor
Interior Sliding Doors			100	Arch	300	Arch	300	Arch	300	Contractor
Interior Folding Doors										
Interior Coiling Doors	8331 OVERHEAD COILING DOORS		200	Arch	300	Arch	300	Arch	300	Contractor
Interior Panel Doors										
Interior Special Function Doors	8390 SPECIAL DOORS		300	Arch	300	Arch	300	Arch	300	Contractor
Interior Access Doors and Panels			NM		100	Arch	200	Arch	300	Contractor

13_Building Information Model Level of Development Matrix

The purpose of the Level of Development (LOD) matrix is to identify who will be modeling which building element to a specific level of detail at a particular point in time for a particular purpose. The LOD matrix is a road map for the team's collective use of Building Information Model in support of the creation of implementation documents for the project, weaving together traditional roles and responsibilities with downstream shop and fabrication documentation. This page is excerpted from a large spreadsheet using the BIMForum format. (*See More Resources for BIMForum web address.*).

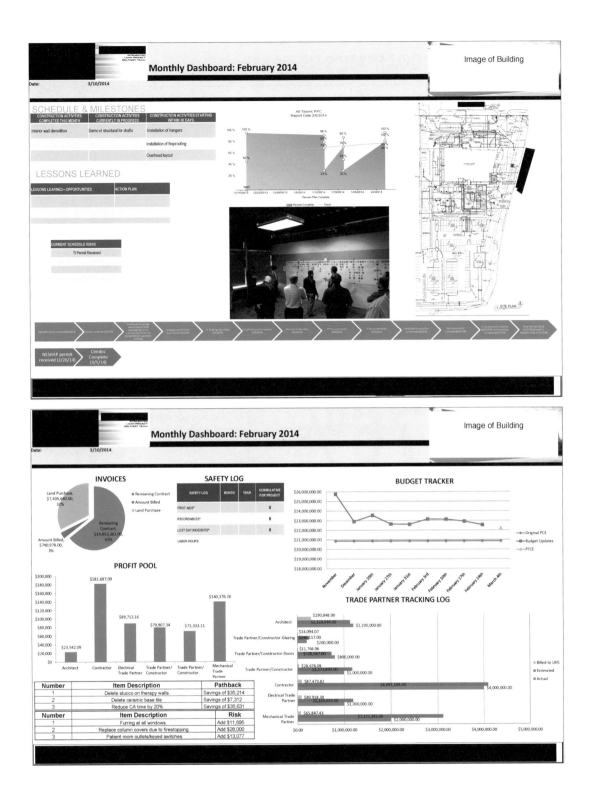

14_Dashboards Example A

Dashboards are a visual management system to track data and metrics that are important to the team and which highlight whether the project is on track and prompts actions. The dashboards on this page are comprehensive, providing a lot of information in two pages; the dashboard on the following page is focused on reliable commitments. Photos of dashboards in the Big Room show how they work as visual management.

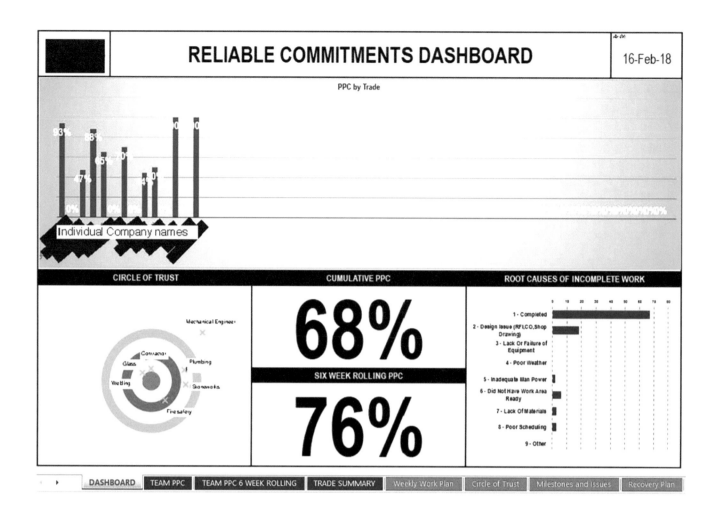

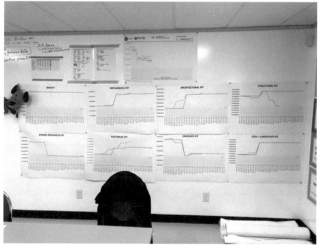

14_Dashboards Example B and Photos of Dashboards in Big Room

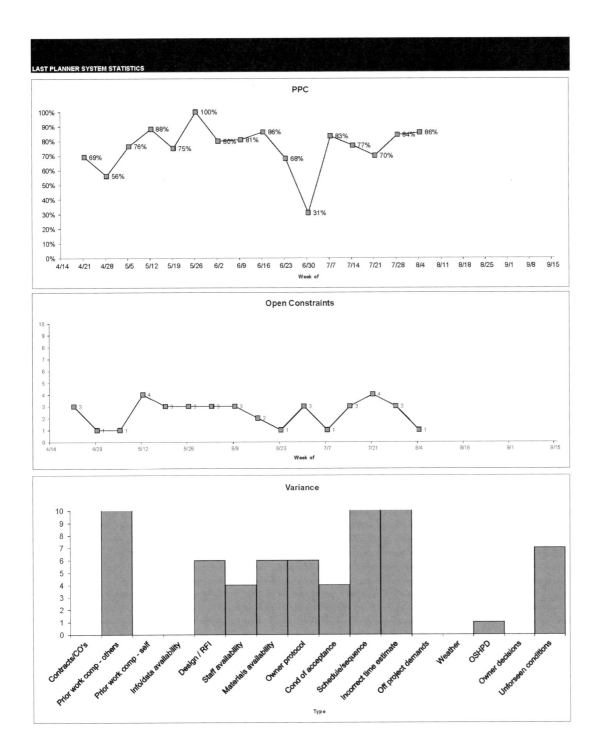

15_Last Planner System

This example shows three data visualizations related to the Last Planner System. At the top is a line graph indicating the IPD team's overall percent plan complete over time (this data is tracked each week with a weekly work plan, such as Appendix 16). Below this is a line graph tracking the number of open constraints over time; this would often be used with a constraint log. (*See Appendix 11 for examples.*) At the bottom is a bar graph showing the number of times that specific reasons contributed to unmet commitments. Together, the data reveals trends in IPD team commitments and impact of obstacles.

Sample courtesy of Paulo Napolitano and Herrero Builders.

1 Contracts/CO's	9 Cond of acceptance	
2 Prior work comp - others	10 Schedule/sequence	
3 Prior work comp - self	11 Incorrect time estimate	
4 Info/data availability	12 Off project demands	
5 Design / RFI	13 Weather	
6 Staff availability	14 OSHPD	
7 Materials availability	15 Owner decisions	
8 Owner protocol	16 Unforseen conditions	

Project   **13236**

Planner

Plan for   **August 4, 2014**

WEEKLY WORK PLAN   PPC **86%**

SCHEDULE ID	REPEAT	ACTIVITY	PERF.	Mon 8/4	Tue 8/5	Wed 8/6	Thu 8/7	Fri 8/8	Sat 8/9	Sun 8/10	DONE? YES NO / CAT.	REASON FOR VARIANCE
Builder												
		soft demo at Nutrition Room for fire cabinet	EM			1	1					
		clean up and turn over tub room	EM				1					
		clean up and turn over soiled utility room	EM				1				2	Prior work comp - others
Special Inspections												
Drywall Subcontractor												
		taping at #5169 and #5171 complete	MM	1								
		taping at med room complete	MM	1	1							
		misc. taping complete	MM	1	1							
Plumbing Subcontractor												
		trim at #5169 and #5171 complete	JS		1	1						
		rough in at Clean Utility complete	JS				1	1			11	Incorrect time estimate

16_Weekly Work Plan During Construction

Typically, a weekly work plan is generated during one week and updated with status during the week following. This is an example of an updated weekly work plan showing the number commitments tracked per week for each responsible party. Commitments met are indicated in green; commitments not met in red, with a white numeral tied to the key of reason codes above. The percent plan complete is tracked in the upper right; this feeds into the uppermost graphic in Appendix 15. The reason codes feeds in the lower two graphics in Appendix 15.

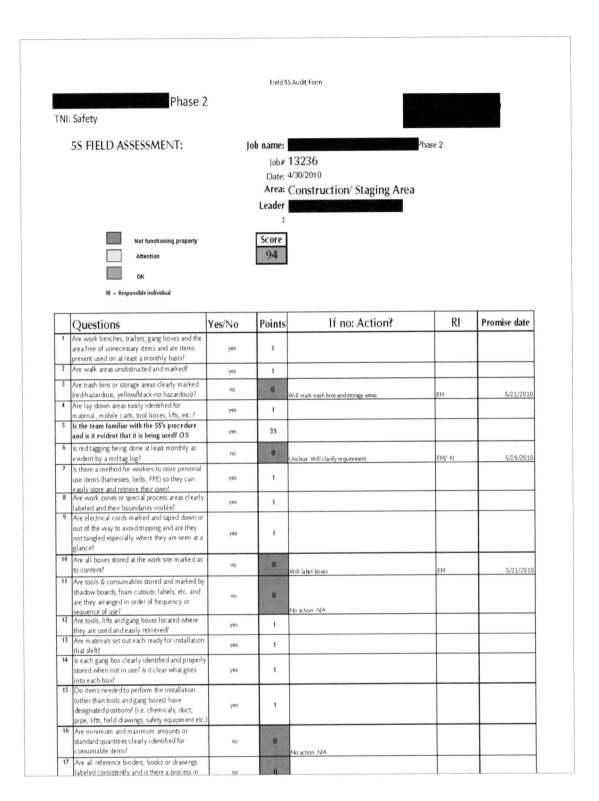

Field 5S Audit Form

████████ Phase 2

TNI: Safety

5S FIELD ASSESSMENT:

Job name: ████████████████ Phase 2

Job# 13236

Date: 4/30/2010

Area: Construction/ Staging Area

Leader ████████████████
:

▨	Not functioning properly
▢	Attention
▨	OK

Score
94

RI = Responsible individual

	Questions	Yes/No	Points	If no: Action?	RI	Promise date
1	Are work benches, trailers, gang boxes and the area free of unnecessary items and are items present used on at least a monthly basis?	yes	1			
2	Are walk areas unobstructed and marked?	yes	1			
3	Are trash bins or storage areas clearly marked (red-hazardous, yellow/black-no hazardous)?	no	0	Will mark trash bins and storage areas	EM	5/21/2010
4	Are lay down areas easily identified for material, mobile carts, tool boxes, lifts, etc.?	yes	1			
5	**Is the team familiar with the 5S's procedure and is it evident that it is being used? (35**	yes	35			
6	Is red tagging being done at least monthly as evident by a red tag log?	no	0	Unclear. Will clarify requirement.	EM/ AI	5/29/2010
7	Is there a method for workers to store personal use items (harnesses, belts, PPE) so they can easily store and retrieve their own?	yes	1			
8	Are work zones or special process areas clearly labeled and their boundaries visible?	yes	1			
9	Are electrical cords marked and taped down or out of the way to avoid tripping and are they not tangled especially where they are seen at a glance?	yes	1			
10	Are all boxes stored at the work site marked as to content?	no	0	Will label boxes	EM	5/21/2010
11	Are tools & consumables stored and marked by shadow boards, foam cutouts, labels, etc. and are they arranged in order of frequency or sequence of use?	no	0	No action. N/A		
12	Are tools, lifts and gang boxes located where they are used and easily retrieved?	yes	1			
13	Are materials set out each ready for installation that shift?	yes	1			
14	Is each gang box clearly identified and properly stored when not in use? Is it clear what goes into each box?	yes	1			
15	Do items needed to perform the installation (other than tools and gang boxes) have designated positions? (i.e. chemicals, duct, pipe, lifts, field drawings, safety equipment etc.)	yes	1			
16	Are minimum and maximum amounts or standard quantities clearly identified for consumable items?	no	0	No action. N/A		
17	Are all reference binders, books or drawings labeled consistently and is there a process in	no	0			

17_Safety Assessment: Leading versus Lagging Indicators

Safety metrics are typically reported as number of days without an accident. However, this is a lagging indicator, since it is tracked after an incident has occurred. This example tracks leading indicators that can help avoid accidents. The 5S field assessment measures potential hazards on site and helps prevent accidents.

APPENDICES

18_ Team Logo

Team culture can be developed in many ways. In this example, team identity and "team first" behavior is supported by a team logo. The project team had seen another IPD team use a team logo and was interested to generate one for themselves. They held an internal design competition. The team members used this logo in their email communications instead of their company logo. In this way, the design-competition process and the resulting logo prioritized identification as a part of the project team over identification as a member of the firm.

	make profit	deliver on schedule	quality	safety	trust	team environment	accountability	open minded	budget	learning	collaboration	operational efficiency	efficiency	aesthetic	supports education	ipd success	tools and processes	innovation	sustainability	roles and responsibilities	quality of life	personal growth	fun	DRIVERS
make profit	■	0	0	0	0	0	0	0	0	0	0	0	0	0	0	0	0	0	0	0	0	0	1	1
deliver on schedule	1	■	0	0	0	0	0	0	1	0	0	0	0	0	0	0	0	0	0	0	1	0	1	4
quality	1	1	■	1	1	0	0	0	1	0	0	1	0	1	1	0	0	0	1	0	1	0	1	11
safety	1	1	1	■	1	0	0	0	1	0	0	nr	1	nr	nr	1	0	0	nr	0	1	0	1	9
trust	1	1	1	0	■	1	0	1	1	1	1	nr	1	nr	1	1	1	1	nr	1	1	1	1	17
team environment	1	1	1	1	1	■	1	1	1	1	1	1	1	1	1	1	1	1	0	0	1	1	1	20
accountability	1	1	1	1	1	0	■	1	1	1	0	1	1	1	1	1	1	1	1	0	1	1	1	19
open minded	1	1	1	1	0	0	0	■	1	1	1	1	1	1	1	1	1	1	1	1	1	1	1	19
budget	1	0	0	0	0	0	0	0	■	0	0	0	0	0	0	1	0	0	0	0	0	0	1	3
learning	1	1	1	1	0	0	0	0	1	■	0	1	1	1	1	1	0	1	1	0	1	1	1	15
collaboration	1	1	1	1	0	0	1	0	1	1	■	1	1	1	1	1	1	1	0	1	1	1	1	18
operational efficiency	1	0	0	nr	nr	0	0	0	1	0	0	■	0	1	1	0	0	0	0	1	nr	1	1	5
efficiency	1	1	1	0	0	0	0	0	1	0	0	1	■	1	1	1	0	0	0	0	1	1	1	11
aesthetic	1	1	0	nr	nr	0	0	0	1	0	0	1	0	■	1	1	0	1	0	1	nr	0	1	8
supports education	0	0	0	nr	nr	0	0	0	1	0	0	0	0	0	■	1	0	0	0	0	1	0	0	3
ipd success	1	1	1	0	0	0	0	0	0	0	0	0	0	0	0	■	0	0	0	0	0	0	0	3
tools and processes	1	1	1	1	0	0	0	0	1	1	0	1	1	1	1	1	■	1	1	1	1	1	1	12
innovation	1	1	1	1	0	0	0	0	1	0	0	1	1	1	1	1	1	■	1	1	1	1	1	12
sustainability	1	1	0	nr	nr	0	0	0	1	0	0	1	1	0	1	1	0	0	■	0	1	0	1	7
roles and responsibilities	1	1	1	1	0	1	1	0	1	1	1	1	1	1	1	0	0	1	1	■	1	1	1	15
quality of life	1	1	1	0	0	0	0	0	1	0	0	0	0	0	0	1	0	0	0	0	■	0	1	5
personal growth	1	1	1	1	0	0	0	0	1	0	0	nr	0	nr	1	1	0	0	1	0	1	■	1	8
fun	1	1	1	0	0	0	0	0	0	0	0	0	0	0	1	1	0	0	0	0	0	0	■	5
DRIVEN	17	14	11	8	3	1	2	3	16	6	3	11	10	9	13	15	6	6	9	4	16	9	17	

Team determined that these equally **drive** & are **driven** by one another:
* Collaboration & Communication
* Safety & Schedule
* Safety & Trust

Unrelated
Equally drive/driven

19_ IPD Team Drivers

This matrix recorded IPD team member opinions on the factors that they believe drive success on a project. The top row indicates the list of project outcomes, such as "make a profit" and "collaboration." The far-left column notes the same list of items, but these represent drivers that lead to a particular outcome listed on the top row. The tally at the bottom shows the total number of drivers that help to achieve a specific outcome. In the drivers column of the matrix (far right), the green boxes indicate what the team believes were the most important drivers: the items the team said actually drove the outcomes of the project. In this case, team environment, accountability, open-mindedness, collaboration, and trust were the highest-rated drivers. The lowest-rated driver was making a profit. In other words, the team viewed collaborative behaviors, communication, and culture as the key project drivers. Seeing the importance of these drivers then helped the team determine what to use as Key Performance Indicator measures. The results documented in this matrix shows that while most project teams track schedule and budget, these are less important indicators in determining whether the project will achieve a successful outcome.

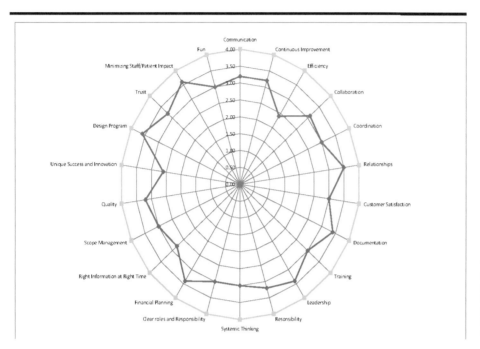

		TNI	SCORE March	previous month
1		Communication	3.20	4.00
2		Continuous Improvement	3.20	4.00
3		Efficiency	2.40	4.00
4	Qualitative	Collaboration	3.10	4.00
5		Coordination	3.00	4.00
6		Relationships	3.50	4.00
7		Customer Satisfaction	3.00	4.00
8		Documentation	3.40	4.00
9		Training	3.00	4.00
10		Leadership	3.40	4.00
11		Resonsibility	3.20	4.00
12		Systemic Thinking	3.00	4.00
13		Clear roles and Responsibility	3.00	4.00
14		Financial Planning	3.40	4.00
15		Right Information at Right Time	2.80	4.00
16		Scope Management	3.00	4.00
17		Quality	3.20	4.00
18		Unique Success and Innovation	2.60	4.00
19		Design Program	3.60	4.00
20		Trust	3.20	4.00
21		Minimizing Staff/Patient Impact	3.60	4.00
22		Fun	3.00	4.00
23	Quantitative	Cash Management		
24		Budget	See Attachments	
25		Safety		

20_Relationship Spider Graph

If the team can agree that aspects of team behavior, communication, and culture are key drivers for success, then these can be mapped on a spider graph. This sample shows the results of a team assessment survey during an IPD project. The results are shown visually in a spider graph, as well as in a table with the aggregated responses and average score per driver. (*See Appendix 19.*)

Sample courtesy of Paulo Napolitano and Herrero Builders.

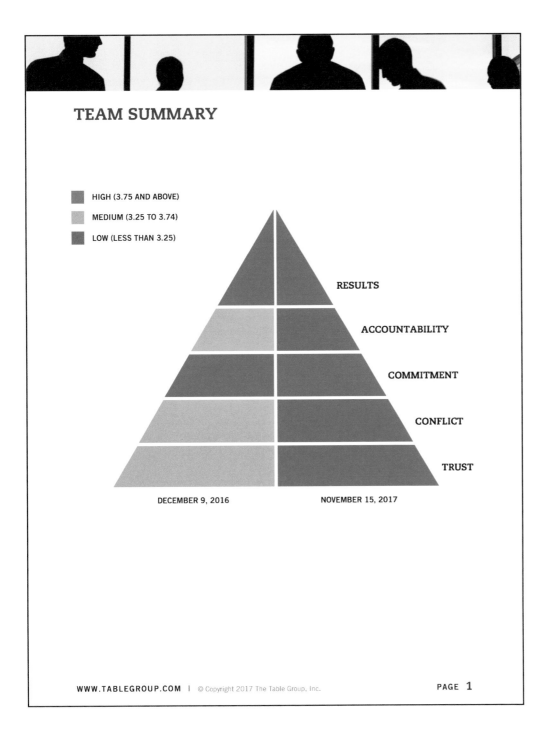

TEAM SUMMARY

HIGH (3.75 AND ABOVE)
MEDIUM (3.25 TO 3.74)
LOW (LESS THAN 3.25)

RESULTS
ACCOUNTABILITY
COMMITMENT
CONFLICT
TRUST

DECEMBER 9, 2016   NOVEMBER 15, 2017

PAGE **1**

21_Tracking, Visualizing, and Improving Team Culture

A forty-person IPD team took two forty-five-minute questionnaires, one in November 2016 and one eleven months later. Results from their responses were aggregated into this visualization. The pyramid shows that in November, the team was strongest in getting results and making commitments; however, managing conflict and mutual trust were weak points. The team analyzed the yellow areas of the pyramid (where they scored poorly), and as a team discussed and put into place countermeasures to address the problems. When the team repeated this exercise in 2017, they found that their team health had improved and the yellow areas of the pyramid were now green.

Sample courtesy of Patrick Lencioni and The Table Group.

APPENDICES

22_Project Financial Update

Periodic financial updates can be complex. This example is too small to read but provides a sense of the level of detail needed for the IPD team to track costs. There are multiple tabs (not all shown here) with target value, actual costs, projected costs, and milestone payouts. They are tracked both by time and as discrete work elements. This dynamic document would be updated every month for review. Therefore, some columns show changes that change over the course of the job.

▉▉ Interiors (Drywall Group)
▉▉▉▉▉▉▉: IFOA Cost Reporting Meeting

Production Conditions	Estimated Hours	Committed Hours	Forecasted Hours	Saved Hours	Cost Savings
Layout	915	806	0	109	$10,838
Full Height Framing	4,247	3,033	165	1,014	$91,909
Ceiling Height Framing	1,690	1,390	90	210	$19,034
Shaftwall Framing	597	526	150	-79	-$7,161
Soffit Framing	1,581	1,166	458	-43	-$3,898
Ceiling Framing	562	502	116	-56	-$5,076
End Caps	151	133	18	0	$0
Wall Insulation	699	876	0	-177	-$16,043
Acoustic Caulking	1,501	1,234	100	167	$15,137
Wall Backing	1,492	1,839	0	-347	-$31,452
Drywall Installation	7,670	7,226	571	-127	-$11,511
Reveals	66	66	0	0	$0
Stocking	572	572	109	0	$0
Clean Up	2,793	2,206	587	0	$0
Fire Caulking	260	260	0	0	$0
Fire Spray	214	51	32	131	$11,874
Fire Taping	1,184	766	155	263	$23,838
Level 4 Finishes	5,922	3,184	742	1,996	$180,917
Finish Trims	1,167	593	263	311	$28,189
Level 5 Finishes	67	123	0	-56	-$5,076
Taping Clean Up	1,001	560	300	141	$12,780
Materials	$1,243,932	$848,604	$220,000		$175,328
Contract Totals	34351	27112	3856	3457	$489,629
Changes to Date (See IPD Contingency Log)					($296,648)
Project Total Savings					**$192,981**

23_Drywall Cost Projections

This is one of a set of three appendices (23, 24, 25) that, together, drive team member accountability and show how each IPD team member contributes to the financial health of the project. This document tracks drywall costs. List of activities, estimated hours, committed (completed) hours, and expected (forecasted) are also listed. Saved hours and cost savings are on the rightmost columns; positive numbers show savings when there are fewer hours than estimated; negative are hours/costs that exceeded the estimate. For this trade partner, a blended rate was used to calculate costs based on hours. The variation of hourly rates for different personnel was not important for this level of information. Materials were tracked separately from labor since these costs are relatively easy to predict. All IPD team members were expected to track their financial progress in their own way, which was unique to the trade or design partner. Information from all IPD team members is aggregated into Appendix 24.

Project Name

Monthly Projected vs. Actual

Updated: 10/23/12
Actuals thru Sept 2012

2012	Jan		Feb		Mar	
WP 1 - Below Grade Parking - CA	Projected	Actual	Projected	Actual	Projected	Actual
Architect	$ 30,340	$ 39,211	$ 30,340	$ 28,534	$ 30,340	$ 25,246
Mechanical Engineer	$ -	$ -	$ -	$ -	$ -	$ -
Electrical Engineer	$ 578	$ 577	$ 1,256	$ 2,472	$ 1,256	$ 165
Structural Engineer	$ 593	$ -	$ 593	$ -	$ 469	
Low Voltage Design	$ -	$ -	$ 1,535	$ -	$ 1,535	
	$ 31,511	$ 39,787	$ 33,724	$ 31,006	$ 33,600	$ 25,410
		26%		-8%		-24%

	Jan		Feb		Mar	
WP 2 - Core and Shell	Projected	Actual	Projected	Actual	Projected	Actual
Architect	$ 74,238	$ 98,967	$ 22,016	$ 49,959	$ 19,492	$ 68,816
Mechanical Engineer	$ 3,780	$ -	$ -	$ -	$ 659	$ -
Electrical Engineer	$ 10,754	$ 10,014	$ 2,512	$ 2,575	$ 1,916	$ 1,684
Structural Engineer	$ 1,778	$ 37,056	$ 1,778	$ 5,620	$ 1,407	
Low Voltage Design	$ 1,875	$ -	$ 1,875	$ -	$ 1,875	
	$ 92,425	$ 146,057	$ 28,181	$ 58,155	$ 25,349	$ 70,500
		58%		106%		178%

	Jan		Feb		Mar	
WP 3 - Tenant Improvements	Projected	Actual	Projected	Actual	Projected	Actual
Architect	$ 299,098	$ 279,223	$ 311,178	$ 323,040	$ 165,096	$ 190,067
Mechanical Engineer	$ 50,225	$ 51,732	$ 3,650	$ 3,646	$ 5,892	$ -
Electrical Engineer	$ 43,378	$ 39,001	$ 19,476	$ 29,540	$ 21,300	$ 13,702
Structural Engineer	$ 17,382	$ -	$ 17,382	$ 7,347	$ 13,756	
Low Voltage Design	$ 14,250	$ 12,183	$ 16,940	$ 16,919	$ 13,550	$ 8,523
	$ 424,333	$ 382,138	$ 370,626	$ 380,492	$ 219,594	$ 212,291
		-10%		3%		-3%

	Jan		Feb		Mar	
WP 4 - 401 Site and Parking Structure	Projected	Actual	Projected	Actual	Projected	Actual
Architect	$ 37,484	$ 37,458	$ 82,188	$ 79,056	$ 83,036	$ 110,649
Mechanical Engineer	$ 125	$ 129	$ 650	$ 644	$ 2,802	$ 1,936
Electrical Engineer	$ 3,761	$ 3,610	$ 7,536	$ 3,633	$ 5,152	$ 17,800
Structural Engineer	$ 15,352	$ 18,984	$ 18,400	$ 27,798	$ 13,456	
Low Voltage Design	$ 1,360	$ -	$ 2,315	$ 1,020	$ 4,355	$ 3,328
	$ 58,082	$ 60,181	$ 111,089	$ 112,150	$ 108,801	$ 133,712
		4%		1%		23%

	Jan		Feb		Mar	
All Work Packages	Projected	Actual	Projected	Actual	Projected	Actual
Architect	$ 441,160	$ 454,878	$ 445,722	$ 480,589	$ 297,964	$ 394,777
		3%		8%		32%
Mechanical Engineer	$ 54,130	$ 51,861	$ 4,300	$ 4,290	$ 9,353	$ 1,936
		-4%		0%		-79%
Electrical Engineer	$ 58,471	$ 53,202	$ 30,780	$ 38,221	$ 29,624	$ 33,350
		-9%		24%		13%
Structural Engineer	$ 35,104	$ 56,040	$ 38,152	$ 40,765	$ 29,088	$ -
		60%		7%		-100%
Low Voltage Design	$ 17,485	$ 12,183	$ 24,665	$ 17,939	$ 21,315	$ 11,850
		-30%		-27%		-44%
	$ 606,350	$ 628,163	$ 543,619	$ 581,803	$ 387,344	$ 441,913
		4%		7%		14%

24_Design Team Monthly Projected versus Actual: Aggregating the Data

This document shows the monthly projected costs as compared to actual costs for multiple firms of the design team (architect, engineer, consultants). The vertical column in yellow shows the actual monthly cost, and the column in white shows the projected monthly cost. To develop this document, each company completed its own detailed report (*see Appendix 23 for an example of a trade partner company's calculations*), and those reports were aggregated to create this document. This project divided the work into three separate work packages, and cost information is tracked for each. Total costs are listed in the bottom section.

This document would be used by the owner to monitor cost trends of both individual firms and the overall design team. The owner and IPD team can easily see if a firm is over or under their cost projections each month. Changes to the expected hours form the basis for discussion on issues that affected the change. In some cases, schedule may have shifted to stagger costs to occur at different times than anticipated. This tool drives accountability and conversations about current and anticipated workflow with enough time to use countermeasures if negative trends are noted.

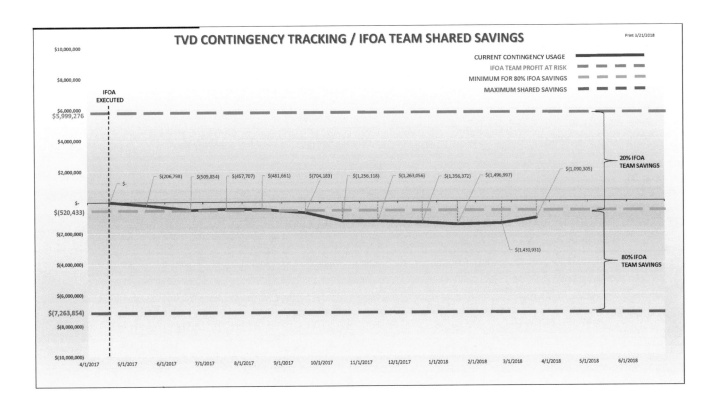

25_ Visualizing Contingency Tracking

This graph visually tracks the cost of a project over time in relationship to the financial terms set up by the IPD agreement. At the center of the graph is a thin black line showing the original cost estimate of the work without the contingency. The zone between the dotted red line at the top and the dotted blue line in the middle indicates the zone where the IPD team will receive 20% of any savings as profit. Between the blue and green, the team will make 80% of savings as profit. If the expenses exceed the red line, the IPD team will not have any profit. If costs are below the green line, the additional savings are kept by the owner. The purple line is the key information tracked by the team—it indicates the actual cost of the project. This information is derived from information similar to that shown in Appendix 24. All of this data can feed into a dashboard.

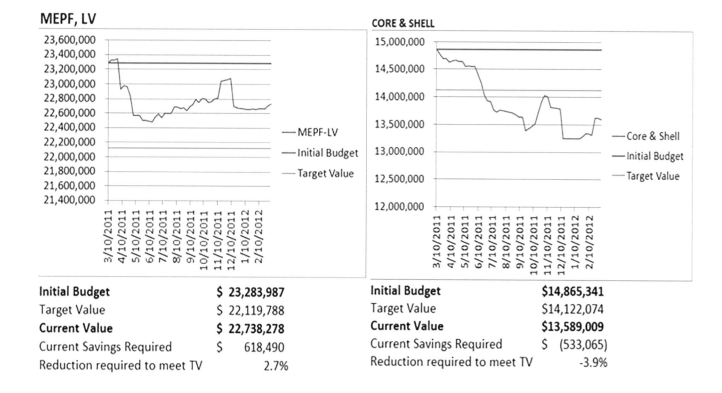

MEPF, LV		CORE & SHELL	
Initial Budget	$ 23,283,987	Initial Budget	$14,865,341
Target Value	$ 22,119,788	Target Value	$14,122,074
Current Value	**$ 22,738,278**	**Current Value**	**$13,589,009**
Current Savings Required	$ 618,490	Current Savings Required	$ (533,065)
Reduction required to meet TV	2.7%	Reduction required to meet TV	-3.9%

26_Cluster Group Tracking in Target Value Design

Project implementation teams (PITs) are formed by the IPD team to develop work related to specific elements, tasks, or areas. They are responsible for tracking the project costs of their work against a subtarget amount that contributes to the total target cost. In target value design, PITs are tracked individually, with the understanding that their work is complementary—one PIT might miss its target because they made a decision that led to a savings for another PIT. This is especially valuable if the saving to one exceeds the overrun for another. The blue line shows monthly status of each PIT relative to their subtarget cost. The top red line indicates the initial budget, and the bottom green line indicates the target. You can see that the mechanical/electrical/plumbing/fire PIT did not achieve its target; however, the core-and-shell PIT was well below their target. All of this data can feed into a dashboard.

BUSINESS OH CALCULATOR FOR IPD PROJECTS

Company: Joe's Roofing
Scope of Work: Roofing
Date: April 9 2015
Project: Building 1

	2014		2013		2012	
REGIONAL YEAR CONTRACT REVENUE	20,000,000		20,000,000		20,000,000	
DIRECT JOB COSTS						
- Field Labour and benefits	1,093,943		1,093,943		1,093,943	
- Materials, vehicles and other	2,441,854		2,441,854		2,441,854	
- Subcontractors	14,030,018		14,030,018		14,030,018	
GENERAL AND ADMINISTRATIVE EXPENSES (OFFICE LOCATION)						
IPD Excluded OH Costs						
- Advertising and promotion	49,416		49,416		49,416	
- Billable Staff (All PMs/Estimators, see list below)	600,000		600,000		600,000	
- Business taxes	493		493		493	
- Donations	6,697		6,697		6,697	
- Legal costs	6,970		6,970		6,970	
- Salary Bonuses	86,717		86,717		86,717	
Amortization & bank charges	62,427		62,427		62,427	
Computer support and maintenance	26,136		26,136		26,136	
Head office corporate Staff (Non Billable)	100,000		100,000		100,000	
Regional corporate Staff (Non Billable)	100,000		100,000		100,000	
Dues and memberships	4,279		4,279		4,279	
Insurance	6,195		6,195		6,195	
Interest	23,458		23,458		23,458	
Office lease, utilities & maintenance	66,915		66,915		66,915	
Professional fees	18,211		18,211		18,211	
Recruiting	22,445		22,445		22,445	
Safety	29,460		29,460		29,460	
Small Tools and Equipment	80,000		80,000		80,000	
Training	21,785		21,785		21,785	
Travel	5,215		5,215		5,215	
TOTAL OH	1,316,822		1,316,822		1,316,822	
Deduct IPD Excluded OH Costs	(750,295)		(750,295)		(750,295)	
TOTAL IPD OH	566,527	3.23%	566,527	3.23%	566,527	3.23%
COST	17,565,816		17,565,816		17,565,816	
PROFIT	1,117,362.4	6.4%	1,117,362.4	6.4%	1,117,362.4	6.4%
TOTAL OH	1,316,822	7.5%	1,316,822	7.5%	1,316,822	7.5%
TOTAL YEARLY REVENUE	20,000,000	100.0%	20,000,000	100.0%	20,000,000	100.0%

AVERAGE 3 YEAR IPD OH   3.23%

List Office Billable Staff To Project
Estimator
Project Manager
Draftsman
Co-ordinator
Project Assistant
Project Director
Project Accountant
Preconstruction Directors
Other
Other
Other
Other
Other

27_Calculating Overhead Costs for Builder

The cost calculators in Appendix 27 and Appendix 28 are from a Canadian contractor. The purpose of these spreadsheets is to foster a conversation to allow the IPD team to establish billable rates for each member, leading to rates that can be used to project costs. Some projects use these templates as part of the request for proposal or request for qualifications process, requesting candidate firms to complete the information in their response.

Cost is broken into three areas: direct cost, indirect cost, and overhead. After each member completes their spreadsheets, the IPD team should have the hourly rates, clearly indicating direct, indirect, and overhead costs, while the profit expectation remains completely separate.

IPD Architect OPERATIONAL COSTS - HOURLY CHARGEABLE COST CALCULATOR

COMPANY NAME:

DATE:

EMPLOYEE CLASS	Hourly Billable Rate	Base Hourly Wage Rate	Direct Personnel Expense Rate	Direct Personnel Expense Multiplier	Indirect Cost & Overhead Rate	Indirect Cost & Overhead Multiplier
	$	$	$		$	
ARCHITECTURAL						
Principal [name}	$101.12	$ 72.12	$ 12.04	0.17	$ 16.96	0.20
Architect [name]	$ 73.07	$ 48.08	$ 8.03	0.17	$ 16.96	0.30
Intern Architect [name]	$ 50.62	$ 28.85	$ 4.82	0.17	$ 16.96	0.50
Technologist [name]	$ 67.46	$ 43.27	$ 7.23	0.17	$ 16.96	0.34

	Direct Personnel Expense Rate					
	Direct Personnel Expense Rate	CPP	EI	WCB	Vacation	Stat. Holiday
	$	0.06%	2.63%	2.00%	8.00%	4.00%
ARCHITECTURAL						
Principal [name}	$ 12.04	$ 0.04	$ 1.90	$ 1.44	$ 5.77	$ 2.88
Architect [name]	$ 8.03	$ 0.03	$ 1.27	$ 0.96	$ 3.85	$ 1.92
Intern Architect [name]	$ 4.82	$ 0.02	$ 0.76	$ 0.58	$ 2.31	$ 1.15
Technologist [name]	$ 7.23	$ 0.03	$ 1.14	$ 0.87	$ 3.46	$ 1.73

	Indirect Costs & Overhead								
	Indirect Costs & Overhead Rate	Rent Utilities Maint	Telephone Internet	Depreciation	Licences/Permits	P. Ed and Seminars	Equipment Lease Repairs and	Professional Fees	Insurance	Corporate Services
	$	$ 3.20	$ 0.48	$ 1.20	$ 0.04	$ 0.40	$ 0.24	$ 1.00	$ 2.00	$ 7.20
ARCHITECTURAL										
Principal [name}	$ 16.96	$ 3.20	$ 0.48	$ 1.20	$ 0.04	$ 0.40	$ 0.24	$ 1.00	$ 2.00	$ 7.20
Architect [name]	$ 16.96	$ 3.20	$ 0.48	$ 1.20	$ 0.04	$ 0.40	$ 0.24	$ 1.00	$ 2.00	$ 7.20
Intern Architect [name]	$ 16.96	$ 3.20	$ 0.48	$ 1.20	$ 0.04	$ 0.40	$ 0.24	$ 1.00	$ 2.00	$ 7.20
Technologist [name]	$ 16.96	$ 3.20	$ 0.48	$ 1.20	$ 0.04	$ 0.40	$ 0.24	$ 1.00	$ 2.00	$ 7.20

NOTES:

1	Basis of Calculation:	Chargeable Rates use the US Federal Acquisition Regulation (FAR) 31.201 - 205 as a basis of inclusions and exclusions in the chargeable rate.
2	Payroll Burden:	Payroll Burden includes: CPP, EI, WCB, Vacation, Stat Holidays, Health and Wellness
3	Overhead Burden:	Rent, Utilities, Telephone, Internet, Depreciation, Licences/Permits, Professional Education (for registered architects), Professional Membership Dues,

☐ Manual Entry required

☐ Calculated Field (see formula)

Corporate Services Include reception, Office Expenses, Meals, Automotive, Postage, Courier, Travel, Interest Charges,

28_ Calculating Overhead Costs for Designers

Note that overhead can be calculated in different ways. The method used by design teams to calculate overhead can be vastly different from how builders calculate it. While there is no right (or easy) way to calculate overhead, there must always be a deliberate and transparent approach to the calculation.

Appendix 27 shows overhead calculations for a builder that uses a blended rate for all staff from that company. Appendix 28 is an example of the cost calculator used for architects, with different factors and priorities than for the builder.

CORE GROUP

Core Group and project manager at the May 2nd workshop. From left to right: Howard Ashcraft, Renée Cheng, Markku Allison, James Pease, Laura Osburn, and Sue Klawans.

Howard Ashcraft, Partner, Hanson Bridgett LLP, San Francisco, USA; Adjunct Professor Civil and Environmental Engineering, Stanford University

Howard Ashcraft has led in the development and use of Integrated Project Delivery and Building Information Modeling in the United States, Canada and abroad. Over the past decade, his team has structured over 140 pure IPD projects and many other highly-integrated projects. He co-authored the AIACC's Integrated Project Delivery: A Working Definition, the AIA's IPD Guide, Integrating Project Delivery (Wiley 2017) and chaired subcommittees for the National Building Information Modeling Standard (NBIMS). A partner in the San Francisco law firm of Hanson Bridgett, he is an elected a Fellow of the American College of Construction Lawyers, an Honourary Fellow of the Canadian College of Construction Lawyers and is an Honorary Member of AIA California Council. In addition to his practice, he is an Adjunct Professor of Civil and Environmental Engineering at Stanford University.

Markku Allison, AIA, Director of Engagement and Innovation, Chandos Construction, Alberta, Canada; President, Integrated Project Delivery Alliance, Canada

Markku has over 35 years experience both as an award-winning designer and as a thought-leader on design and construction industry transformation issues. His background as a practice owner and industry subject matter expert with strong relationships across disciplines and organizations uniquely positions him to assist in shaping responsive strategies to forces driving change in business and culture today. In his current role, Markku heads up a portfolio that includes Chandos' IPD, Lean, Sustainability, BIM, and Marketing initiatives. He serves as President of the Integrated Project Delivery Alliance (www.ipda.ca) in Canada which has published several influential IPD research studies and provides IPD training for industry. Markku held positions at the American Institute of Architects, where he played a key role in developing AIA's "Integrated Project Delivery: A Guide".

Renée Cheng, FAIA, DPACSA, Professor, School of Architecture, University of Minnesota; Director, M.S. of Research Practices and Consortium for Research Practices

Renée is a nationally renowned educator, in 2019, she will change roles and institutions as Dean of the College of Built Environments at the University of Washington. Cheng pioneered research surrounding the intersection of design and emerging technologies including work on industry adoption of IPD, BIM and Lean. Her research focuses on collaboration, innovation and change, especially how they are fostered by equitable and inclusive practices. Her case studies include one of the first studies on IPD for AIA and the extensive, "Motivation and Means: How IPD and Lean Succeed", co-sponsored by IPDA and LCI. She is active with the American Institute of Architecture, served as the inaugural Chair of the Lean Construction Institute Research Committee and currently serves on the LCI Board of Directors.

Sue Klawans, Consultant, AGC Public/Private Industry Advisory Council Chair

With over 30 years' experience in the industry, Sue is a recognized leader in the industry, sought after to speak on high-performance teams, lean, technology and process innovation, productivity and prefabrication, and metrics and key performance indicators. She combines a background in planning, design, and construction with proven experience and business results as a senior executive and Lean strategist. Sue has managed and facilitated both long-term and annual strategic planning, resulting in advancements in employee development, profitability, and risk management. She also devised and implemented a unique, multi-dimensional Lean program focused on elevating people and teams to achieve breakthroughs and reach new levels of capability. Sue participates in multiple organizations including the Construction Owners Association of America Industry Liaison Committee, BIMForum Strategic Advisory Council, Construction Quality Executives Council Board of Governors, and National Institute of Building Sciences' Off-Site Construction Council Board.

James Pease — IPD Owner's Representative

James Pease is an expert in the set up and structure of large, complex capital projects using Lean and Integrated Project Delivery to drive highly reliable results. He has negotiated IPD contracts and delivered over $650M in complex healthcare projects as an Owner's Representative with multiparty contracts, aligned team incentives and collaborative delivery models. He is the executive editor of the IPD focused website leanipd.com, an LCI NorCal Core Team Member and Co-founder of the COAA California Chapter. James has a BS in Management Science from UC San Diego and is a CA Licensed General Contractor.

FUNDERS

We are grateful for the generosity of our funders for supporting the creation of this guide. This guide was funded by the Pankow Foundation in a grant titled Integrated Project Delivery (IPD) Practitioners Guide (CPF RGA #04-17), distributed to and managed by the University of Washington. Matching funds were provided by Array Architects, Boldt, Cammisa + Wipf, CH2M (now Jacobs), Chandos Construction, Charles Pankow Builders, Clark Construction, DLR Group, DPR Construction, Ferguson Corporation, Gilbane, Gillam Group, Group 2 Architecture Interior Design, Procter & Gamble, Robins & Morton, Rosendin Electric, Southland Industries, and Whiting-Turner.

Made in the USA
Columbia, SC
15 December 2023